Grounds for Knowledge

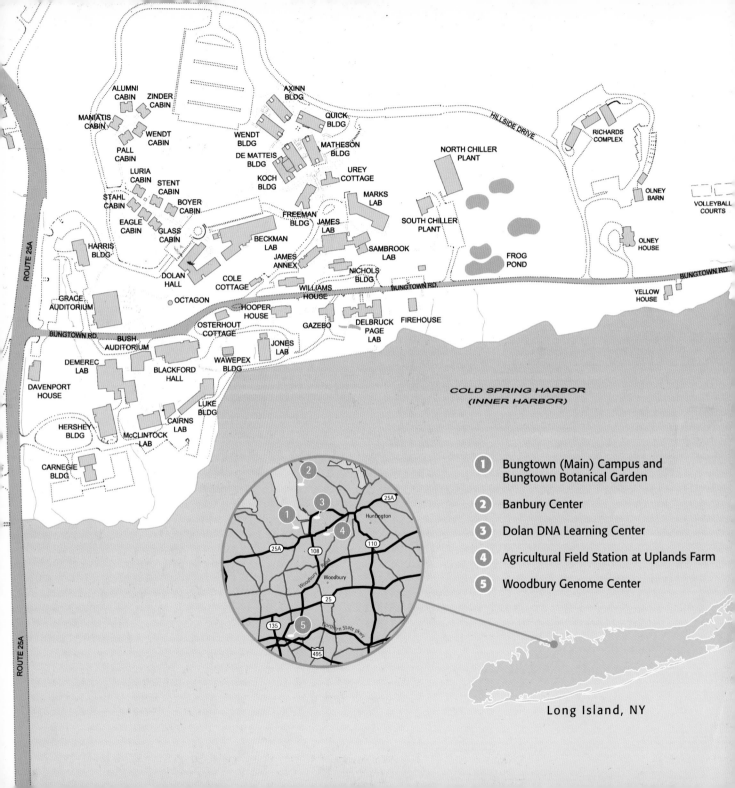

ALUMNI CABIN
ZINDER CABIN
MANIATIS CABIN
WENDT CABIN
PALL CABIN
LURIA CABIN
STENT CABIN
STAHL CABIN
BOYER CABIN
EAGLE CABIN
GLASS CABIN
HARRIS BLDG

AXINN BLDG
QUICK BLDG
WENDT BLDG
MATHESON BLDG
DE MATTEIS BLDG
KOCH BLDG
UREY COTTAGE
MARKS LAB
FREEMAN BLDG
JAMES LAB
BECKMAN LAB
JAMES ANNEX
SAMBROOK LAB
NICHOLS BLDG
DOLAN HALL
COLE COTTAGE
WILLIAMS HOUSE

HILLSIDE DRIVE

RICHARDS COMPLEX
OLNEY BARN
VOLLEYBALL COURTS
NORTH CHILLER PLANT
SOUTH CHILLER PLANT
OLNEY HOUSE
FROG POND
BUNGTOWN RD.
YELLOW HOUSE

ROUTE 25A

GRACE AUDITORIUM
OCTAGON
HOOPER HOUSE
OSTERHOUT COTTAGE
BUNGTOWN RD
BUSH AUDITORIUM
GAZEBO
DELBRUCK PAGE LAB
FIREHOUSE
DEMEREC LAB
BLACKFORD HALL
JONES LAB
WAWEPEX BLDG
DAVENPORT HOUSE
LUKE BLDG
CAIRNS LAB
HERSHEY BLDG
McCLINTOCK LAB
CARNEGIE BLDG

COLD SPRING HARBOR
(INNER HARBOR)

ROUTE 25A

1. Bungtown (Main) Campus and Bungtown Botanical Garden
2. Banbury Center
3. Dolan DNA Learning Center
4. Agricultural Field Station at Uplands Farm
5. Woodbury Genome Center

25A
Huntington
110
25A
108
Woodbury
Woodbury Road
25
135
495
Northern State pkwy

Long Island, NY

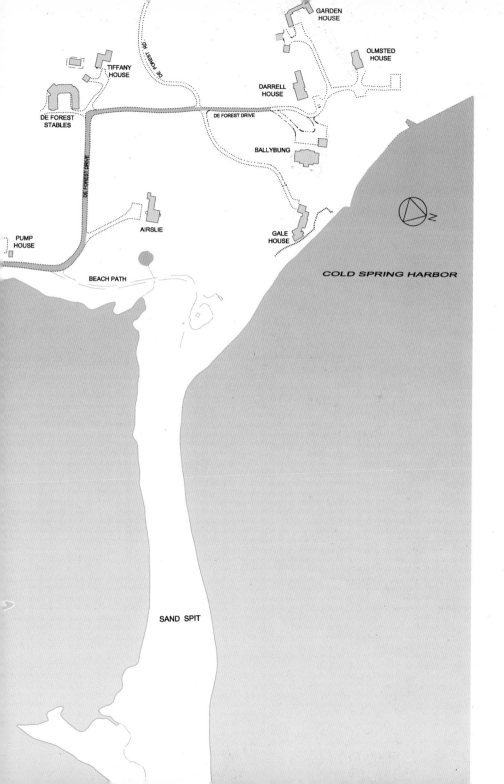

GARDEN HOUSE

OLMSTED HOUSE

TIFFANY HOUSE

DE FOREST RD

DARRELL HOUSE

DE FOREST STABLES

DE FOREST DRIVE

DE FOREST DRIVE

BALLYBUNG

AIRSLIE

PUMP HOUSE

GALE HOUSE

N

BEACH PATH

COLD SPRING HARBOR

SAND SPIT

Bungtown Campus Map

Inset:
Cold Spring Harbor
Laboratory Campus
Locations

Cold Spring Harbor Laboratory
1 Bungtown Road
Cold Spring Harbor, NY 11724

•

www.cshl.edu

Cold
Spring
Harbor
Laboratory

Grounds for Knowledge

A GUIDE TO COLD SPRING HARBOR LABORATORY'S LANDSCAPES & BUILDINGS

Introducing the Bungtown Botanical Garden

LANDSCAPE PHOTOGRAPHY BY PETER STAHL

Elizabeth L. Watson

FOREWORD BY VINCENT A. SIMEONE

PREFACE BY JAMES D. WATSON

Cold Spring Harbor Laboratory Press, 2008

© 2008 by Cold Spring Harbor Laboratory Press, Cold Spring Harbor, NY
All rights reserved. Printed in the United States of America
Design by Frederick Dress and Abby Dress
Editing and Production by Rosalie Ink Publications

Front Cover: Bright yellow Ginkgo with Cold Spring Harbor Laboratory's Gale House (l)
and Ballybung, autumn, along the western shore of Cold Spring Harbor.

Back Cover: Jones Laboratory in ice and ivy . . . the Firehouse . . . Flowering Dogwood
in early spring.

Library of Congress Cataloging-in-Publication Data

Watson, Elizabeth L.
  Grounds for knowledge : a guide to Cold Spring Harbor Laboratory's
landscapes & buildings / Elizabeth L. Watson ; landscape photography by
Peter Stahl.
      p. cm.
  Includes bibliographical references and index.
  ISBN 978-0-87969-799-0 (hard cover : alk. paper)
  1. Bungtown Botanical Garden (Cold Spring Harbor, N.Y.)--Guidebooks. 2.
Cold Spring Harbor Laboratory--Guidebooks. 3. Cold Spring Harbor
(N.Y.)--Guidebooks. I. Title.

  QK73.U62B868 2008
  580.73'74721--dc22

                                        2007051534

10 9 8 7 6 5 4 3 2 1

All Cold Spring Harbor Laboratory Press publications may be ordered directly from Cold Spring Harbor
Laboratory Press, 500 Sunnyside Blvd., Woodbury, New York 11797-2924. Phone: 1-800-843-4388 in Con-
tinental U.S. and Canada. All other locations: (516) 422-4100. FAX: (516) 422-4097. E-mail: csh-
press@cshl.edu. For a complete catalog of all Cold Spring Harbor Laboratory Press publications, visit our
World Wide Web Site http://www.cshlpress.com/.

For Faith H. McCurdy

Contents

Illustrations and Maps

Foreword

Trees are without a doubt an integral part of our environment. What would our cultivated landscapes and communities be without trees? Undoubtedly pretty boring. Trees offer unlimited beauty and great function in the landscape. They enhance our quality of life with their stately habit and seasonal interest.

This unique and inspiring volume *Grounds for Knowledge*, presented by Elizabeth L. Watson, is a great tribute to the history and splendor of the Cold Spring Harbor Laboratory. It reminds us how important the landscape around us really is. This world-renowned research center continues to be at the forefront of great science and higher learning but it is also an example of a well-designed landscape with distinctive beauty. My endorsement of this text comes from knowing the author and the facility of which she writes for over fifteen years.

During my first visit to the Cold Spring Harbor Laboratory, I was amazed at the maturity and grace of so many wonderful trees. Some of the most majestic specimens of Horsechestnut, Black Locust, Ginkgo, and Beech can be found sprinkled throughout the property. Among my favorite trees are the immense Amur Corktree (*Phellodendron amurense*) and Kobus Magnolia (*Magnolia kobus*), the largest of their type on Long Island, located near the *Time Spirals* sculpture. Equally impressive is the diversity of plantings throughout the grounds. The way the beautiful architecture and picturesque landscape intermingle and complement each other is a rare and unique combination possessed by few facilities. I can speak from experience having served on the Lab's Tree Committee, which was created by the author with the sole purpose of preserving, enhancing, and promoting trees.

It is easy to appreciate the motivation behind this book once you understand the root of Liz Watson's passion for trees and her appreciation for the landscape. I first met Liz when I was an intern at Planting Fields Arboretum in Oyster Bay, New York. I was always so impressed by her knowledge of historic preservation. Years later she accompanied one of my mentors, Allan M. Armitage, and myself, along with almost thirty other enthusiastic participants, on a tour of great gardens of Eastern Canada. The trip involved visiting more than twenty public and private gardens with the goal of bringing back good ideas for our own gardens.

Afterwards I had the pleasure of having Liz as one of my students in a woody plants class that I teach annually at Farmingdale State University. Her ability to absorb and retain rather complex and comprehensive plant information was remarkable, and this experience seemed to transform her into a bona fide tree lover.

As with everything she does, Liz's love of life and her work is evident in each page of this beautifully crafted text. The devotion and passion that she and her husband James Watson have instilled into the Cold Spring Harbor Laboratory is to be admired. Read on and enjoy!

Vincent A. Simeone, Director
Planting Fields Arboretum State Historic Park

Preface

That trees can be as serenely beautiful as the skyscrapers of New York City and Chicago or as awe-inspiring as the ice-covered Alps above Zermatt only entered my psyche after my marriage in 1968 to Elizabeth Vickery Lewis, the author of this welcome new book. I had just been appointed Director of the land-rich but money-impoverished Cold Spring Harbor Laboratory, located on Long Island's North Shore close to many of the large Gold Coast estates that came into existence between the Civil War and World War II. At Harvard, Liz first became excited by large-scale domestic architecture through lectures by James Ackerman on Palladio. After our sons Rufus and Duncan were born, she followed up her new-found passion for grand architecture by enrolling in the historic preservation program of the Columbia University School of Architecture. Her first courses were still in progress when she persuaded me to broaden the scope of subsequent visits to England. Instead of focusing on London for its theatre and music and Cambridge for its science, we would spend equal time touring the English countryside to visit its many stately homes. Our first such occasion was in 1977 to the west of London. In four days, we excitedly took in the grand excesses of Osterley House in Middlesex, Blenheim Palace in Oxfordshire, and Longleat and Wilton House in Wiltshire.

Then I came away sensing that these elegant architectural triumphs were much more humanized by their surroundings than by the furniture and paintings of their long ago inhabitants. Their gardens and their walls, their lakes and associated grottos, their horses and sheep, and most of all their trees were what truly excited my heart. When they are standing singly as glorious "specimen" trees, interdigitated as collective groups, say, a copse of beeches, scattered irregularly as ashes or oaks along garden paths, or bordering the long allées going from entering gates to the entrance steps, ancient trees raise the human spirit.

Our so instinctively focusing on the trees of the English countryside reflected the Cold Spring Harbor Laboratory's small but spectacular collection of specimen trees on lands to its north that once were part of the more than one hundred acre Henry de Forest estate. As President of the New York Botanic Garden and shrewd investor in Western Railroads, de Forest had the means to mimic England at its best and toward that end, in the early 1900s, hired the famed Olmsted Brothers to redesign his formal garden so its spring elegance had no equal on Long Island.

— ✦ —

. . . the trees of the English countryside reflected the Cold Spring Harbor Laboratory's small but spectacular collection of specimen trees on lands to its north that once were part of the more than one hundred acre Henry de Forest estate.

— ✦ —

Liz and I chose our first trees for the Lab when the dilapidated, early 1800s Osterhout Cottage was rebuilt in 1969 to serve as our first home at the Lab. Two thousand dollars earlier invested by me in the rapidly growing birth control pill juggernaut Synthex Pharmaceuticals was by then worth $30,000, covering reconstruction costs and much-welcomed air conditioning. To our new home's immediate south, we planted a tiny Copper Beech that today is on its way to being a specimen tree. Less wise was our planting a Blue Atlas Cedar next to Bungtown Road. It never had space to grow into magnificence and last year had to be cut down. When Manny Delbrück provided monies needed to change Wawepex from a bedraggled lab to a sixteen-bed dormitory, we planted beside it a small Gingko that today is beginning to stand out as a visual delight. Above it we placed a coniferous Cryptomeria from Japan whose columnar beauty we first appreciated on property near our then summer home on Martha's Vineyard. Further down Bungtown Road, to the north of the Firehouse, we planted a tiny Metasequoia from China. Though now more than fifty feet tall and more than a foot in diameter, only its top is truly alive due to our failure to cut down the not now small trees around it.

After five years in Osterhout, we moved down Bungtown Road to the long ago farmhouse Airslie, with its almost century-old spectacular

Kobus Magnolia and Horsechestnut trees, both of whose lower branches projected secondary roots into the soil. We also inherited a lovely collection of ancient white Flowering Dogwoods that much too soon fell victim to the blight that badly diminished the magnificence of their May blooms. To replace them we began planting their much more disease-resistant Kousa relative from Japan. Near the kitchen door we inserted a small reddish Copper Beech that by now almost rivals in impact the much earlier planted Beeches to the northwest. As the massive Weeping Willows which grew over its spring-fed pond were beginning to show their old age, we complemented them by having two more placed near to them.

Between the once de Forest land and the main collection of Lab-owned buildings was a large field on which sheep had long ago grazed. By the time Liz and I took up residence in Airslie, the field was covered by an impenetrable collection of Celastrus vines and stunted trees that gave protection to the many pheasants emanating from the still existent large Schiff estate on which hunting had occurred. To clear it for use as a potential field for sports, our grounds superintendent Buck Trede and his then small crew employed their trusty bulldozer to spend much of a spring and summer opening long grass vistas and letting Olney House, by then repainted in original colors, be seen from afar.

Our personal involvement in placing trees diminished when the Lab's much expanded scientific and educational efforts necessitated the construction of brand new buildings. How to landscape our 1986 Grace Auditorium was not for amateurs, and we turned for help to George Betzel of Long Island's famed landscape firm Innocenti and Webel. Without him, there would not be Willow Oaks to its north or the elegant Weeping Hemlock and Japanese Maple at its main entrance.

By now the ever-growing size and complexity of the Lab's buildings and maintenance needs have led to increased internal expertise on

> As the massive Weeping Willows which grew over its spring-fed pond were beginning to show their old age, we complemented them by having two more placed near to them.

landscape matters. Our own Danny Miller oversees the current massive reforesting project necessitated by the recent clearing of more than fifteen acres of Oaks, Hickories, and Ashes on lands once used for farming bordering on Moores Hill Road. On this cleared space is now arising a new hillside set of science buildings designed to let the Lab exploit the extraordinary potential for the just completed human DNA sequence letting us understand the molecular essences of cancer and psychiatric disorders. Just to its west will be an upper campus of residential and teaching buildings whose hoped for erection over the next ten to fifteen years will even more indelibly mark us as one of the world's premier academic centers for teaching and research.

Now when Liz and I walk about the Lab grounds or make much longer perambulations about Caumsett, Marshall Field's once grand English-style estate in nearby Lloyd Harbor, we take as much pleasure from jointly identifying trees as long ago as a young man I got from spotting birds with my father. *Grounds for Knowledge* handsomely reflects Liz's studies and skills, and the work of many in honoring the Lab's campuses and the newly designated Bungtown Botanical Garden.

Through trees we look back to our pasts and anticipate our futures.

Jim Watson

Acknowledgements

All of us who enjoy the beauty of the grounds of Cold Spring Harbor Laboratory owe a debt of gratitude to its professional stewards. In particular I would like to single out, in order of my personal acquaintance, Donald Eckels, a University of California at Davis graduate who, as Superintendent of Buildings and Grounds of CSHL in the 1960s, supervised the planting of most all the specimen trees that have now reached maturity. I think of Don just about every time I walk past the looming Insensecedar across the road from the Nichols Building – a tree from the California Rockies that found a new home almost forty years ago on the gentle Long Island slopes of Cold Spring Harbor.

Don was succeeded in 1970 as B & G superintendent by Jack Richards, a local Huntington builder of fine homes who was persuaded to join the CSHL staff to supervise construction of an Annex to provide offices and seminar space for scientists studying cancer in James Lab. Jack's in-house contracting skills were needed because, with the appointment of Jim Watson as Laboratory director, the moment had arrived to quicken the pace of renovating the Lab's rundown physical plant – to find the clues of cancer causation through the brand new field of tumor virology. Jack soon found himself working closely with the talented architects and planners from the Essex, Connecticut, firm known then as Charles Moore Associates, later Moore Grover Harper and today Centerbrook. More often than not, Jack would tinker with their creative, albeit contextual, designs to make sure they could actually be built with the Lab's limited finances. It all worked out very well, for this year is the thirty-fifth in succession of a most productive relationship between the Laboratory and the wonderful professionals headquartered in Essex; it was all the way back in 1973 that we first sat down together – with Charles Willard Moore himself!

Meantime Jack hired plantsman Hans ("Buck") Trede to manage the grounds which were in a perpetual state of upheaval as each old building was renovated, often with additions, and always requiring its immediate

landscape to be renewed. Buck *loves* Junipers, but there wasn't a plant he encountered at CSHL that he couldn't readily identify. (Also, he could pull up poison ivy with his bare hands.) Buck and his crew developed expertise in creative trenching for handling all the latest wiring and cabling requirements for science, plus the perennial problems of all those "cold springs" traversing the Laboratory grounds.

Buck was succeeded by Chris McEvoy, who now reigns as the Lab's resident horticulturalist. Today, Daniel Miller manages the grounds. Danny's background in "tree surgery" has saved many a specimen planting, but his most recent feat has been replanting a native forest in and about the Hillside and future Upper campuses – with a little help from the Crew! He has also personally ordered and "staked" the almost 200 customized tree identification plaques that now indicate to our visitors from near and far that the Bungtown campus of Cold Spring Harbor Laboratory is indeed a botanical garden.

Our "tree man" Danny Miller, as the grounds manager of CSHL –

together with Frank Russo, our "buildings man," who as manager of construction is cheerleader to all of the CSHL in-house trades professionals – both report to Art Brings, the Laboratory's chief facilities officer. Art is "Mr. Can-Do" par excellence; ask any of "his" scientists . . . or neighbors!! Also on Art's team is infrastructure maven Peter Stahl, who just happens to be responsible for the overwhelming majority of the lovely landscape images in this book. Luckily Peter gets to work before most other staffers so his images rarely portray autos (or people for that matter!). The handsome maps in the book have been created by Daniella Cunha Ravn, in-house architect and interior designer at the Lab.

Most of this book was written over the course of two summers in the CSHL Press "Authors' Conference Room" on the second floor of the Carnegie Building. There my suitemate – and No. 1 cheerleader – was Press staffmember Maryliz Dickerson. (I recently learned her husband Bill helped "paste up" *Houses for Science* back in 1991; those were the good ole days.) Each summer I had technical help vis-à-vis the photographic images from editorial assistants of my husband, first Marisa Macari (now pursuing a career in biological anthropology at Oxford, UK) and second, Agnieszka Milczarek (now pursuing a medical degree at Washington University in St. Louis).

Not long after Agnieszka departed for medical school, the Cold Spring Harbor Laboratory Library and Archives were preparing to vacate the Carnegie Building so that it could be readied for the addition of a wing for the study of the history of molecular biology and bio-technology. With one foot out the door, I latched myself onto kindred spirit Linda Swanson, who has been working part time for the executive director of Library and Archives, Mila Pollock, while pursuing her library degree (as I once had) at the Palmer School of nearby Long Island University.

As they say, "Ask a librarian!" I got the OK from Mila to kidnap Linda temporarily, and she has not only surpassed herself in encouraging the photos and their respective credits along, but made good manuscript suggestions – and, has also just completed the requirements for her

MSLIS degree. Thanks are due as well to CSHL's Phil Renna, for trawling the massive digital files of the Media Arts and Visualisation Department for those elusive last minute images, and to Joan Lui for making readily accessible her files of all the digital images submitted in recent years by CSHL's "amateur" photographers for possible inclusion in the Laboratory's beautiful annual fund-raising calendar.

I describe in my Introduction the invaluable horticultural help I received from Vincent A. Simeone and Mrs. Frances Elder, who planted various tree seeds in my vivid imagination – thanks, Vinnie and Franny!

Finally I would like to single out the efforts of John Inglis, PhD, CSHL Press's executive director and publisher, and his staff members Elizabeth Powers, Denise Weiss, and Kaaren Hegquist in seeing this book to completion. And words cannot express my gratitude to John for, in his infinite wisdom, contracting out the editorial and design aspects of the project to Terry Walton, in my opinion the sine qua non of "Editrices," and to gifted designers Fred and Abby Dress.

Writing a book and watching it progress to publication is quite an adventure. For having this opportunity now twice, I thank Jim, the constant companion of nearly all my life's adventures – here, there and everywhere!

Liz Watson

Introduction

*G*rounds for Knowledge was written in the course of professional-level training in the field of ornamental horticulture, specifically in woody plants – trees, shrubs, and vines. It is a book about the "lay of the land" of Cold Spring Harbor Laboratory and the buildings and plants that exist there. The Laboratory was founded in 1890 on a small parcel of land on the North Shore of Long Island, New York, about thirty miles east of Manhattan. Today the historical main campus of the Lab hugs the entire western shore of Cold Spring Harbor's Inner Harbor. Its roughly forty buildings are arrayed immediately west and east of the main north/south thoroughfare, Bungtown Road, which runs approximately parallel to the shoreline from the head of the harbor all the way to the Sand Spit dividing the brackish Inner Harbor from the full-salt Outer Harbor. Comprising about one hundred acres, the Bungtown campus was entered in its entirety onto the National Register of Historic Places in 1994.

Grounds for Knowledge was a long time in the making, and celebrates both buildings and grounds, and their interconnections over time. . . .

Long ago I was part of a team assembled by Charles E. Peterson, renowned preservationist from Philadelphia, to study future possibilities for the William Robertson Coe mansion on the grounds of the Planting Fields Arboretum, Coe's former estate in Oyster Bay, Long Island. It was my internship for the graduate program in Historic Preservation at Columbia University. Perhaps I was hired for the team thanks to the proximity of my home in Cold Spring Harbor, just five miles to the east, as the crow flies!

Later, while serving as a trustee of the Planting Fields Foundation, I managed finally to write my master's thesis for Columbia. In it I compared the early architecture of three seaside laboratories founded at the end of the 19th century: the 1888 Marine Biological Laboratory at Woods Hole, Massachusetts; the 1895 Harpswell Laboratory in Maine

(which later became the Mount Desert Island Biological Laboratory); and the institution that was founded on Long Island in 1890 as the Biological Laboratory of the Brooklyn Institute of Arts and Sciences. Interestingly enough, in 1904, the Carnegie Institution of Washington established a Station for Experimental Evolution immediately adjacent to the Bio Lab, on the western shore of Cold Spring Harbor. It was these two institutions – the Bio Lab and the Carnegie Station – that merged in 1963 to become the Cold Spring Harbor Laboratory.

In honor of the Lab's Centennial in 1990, I recast my master's thesis as a coffee table book, *Houses for Science: A Pictorial History of Cold Spring Harbor Laboratory*. Mainly I relied on the research I had already done in the Lab's Archives. But I also wrote about what I knew from first-hand experience – and participation! —as the wife of Jim Watson, newly appointed director in 1968. The architectural story began with the rescue of several buildings on the Lab's grounds from the state of near dereliction after the Carnegie Institution ended its Cold Spring Harbor operation in 1962. As described in *Houses for Science*, the fifteen-year period of 1968 to 1983 was primarily one of rehabilitation – and even reconstruction – of decrepit, and sometimes condemned, building stock. More often than not, this involved adaptive reuse of structures erected originally for purposes different from those now needed. Major federal grants, a result of a national War on Cancer being proclaimed, provided the necessary overhead funds for most of these renovations.

> The period of 1968 to 1983 was primarily one of rehabilitation – and even reconstruction – of decrepit, and sometimes condemned, building stock.

By contrast, the following decade 1983-1993 saw not only the construction of numerous building additions but also the erection of several brand new facilities, thanks to two major fund-raising efforts: the Second Century Campaign and the Infrastructure Campaign. The historic scope of *Houses for Science*, ostensibly the first one hundred years of Cold Spring Harbor Laboratory, included nearly all of the first quarter century of Jim Watson's leadership as director. In fact, the story stretched back to the industrial heyday of the Cold Spring Harbor community a good half century before the Laboratory's founding in 1890.

Most recently, the Lab's last fifteen years 1993-2008 – the ones not covered by *Houses* – have seen myriad changes. In writing a "sequel" I knew I couldn't just leap in at the year 1993. For reasons which I'll get into below, I've ended up writing a "fieldguide" to *all* the buildings – and to the grounds as well. But here I would like to summarize from my own viewpoint the events of the last fifteen years, which are incorporated in *Grounds* but obviously missing from *Houses*.

During this period, the Lab's Jim Watson, first as president and then as chancellor, and Bruce Stillman, first as director and now as president, have been pulling out all the stops to ensure an exciting future for this place where so much exceptional science has taken place. The highlights of this latest era include: founding the Laboratory's graduate program, the Watson School of Biological Sciences; building two new neuroscience facilities; establishing a purpose-designed genomics campus; and completing enhancements of all Laboratory campuses via purchase of neighboring buildings and renovations/additions to existing facilities. And last, but physically perhaps the most impressive, is construction of a new "Hillside Campus" of six inter-connecting Brain, Cancer, and Informatics research labs. These in turn are stepping stones to an eventual "Upper Campus" that will be the new home for the Watson School.

Since *Grounds* covers not just the buildings but most other elements of the Laboratory landscapes as well, including outdoor sculpture, trees, and water features, let me describe how this came about – especially the trees bit. Let's go back to the arboretum at nearby Planting Fields for a moment. In the years after publication of *Houses for Science* I became very involved there again, this time as a trustee and eventually Planting Fields Foundation president. While indeed I could name the architect and

Since Grounds covers not just the buildings but most other elements of the Laboratory landscapes as well, including outdoor sculpture, trees, and water features, let me describe how this came about . . .

style of all its handsome estate buildings, I was really embarrassed about my inability to name more than half a dozen of its wonderful trees – this despite being in regular contact with them all for over twenty years! I had come to wish I were a tree expert.

However, I did know that Cold Spring Harbor Laboratory had some very special trees of its own, and those I could identify. In fact there is a tradition, at least a century old and quietly continued by Jim, for the planting of "specimen" trees on the lands of which Cold Spring Harbor Laboratory today is steward.

As it turns out, Jim and I had long been in possession of a "List of Woody Plants, Cold Spring Harbor Lab and Adjacent Land to North, Observed May 6, 1975." The author of this memo, Planting Fields horticulturalist G. Elizabeth Lotowycz, had inspected most of the unusual woody plants on her list in the vicinity of the Laboratory director's house where we were then living with our young family. We were aware that some of the biggest specimens she had cited might have been planted during the early 20th century heyday of the Henry W. de Forest estate – the same period as the W. R. Coe estate – of which the grounds of our house had formed a part.

"Woody plants" are defined as plants with woody parts, i.e. trees, shrubs, and vines. Just for the record, here are the distinctions between them, to quote from *Trees of North America: A Guide to Field Identification* (a Golden Field Guide that you ought to own).

> Trees are woody plants at least 15 feet tall at maturity, with a well-developed crown and a single stem, or trunk, at least several inches in diameter. Shrubs, also woody plants, have several stems growing from a clump and are usually smaller than trees. Vines may have woody stems but do not have a distinct crown of upright branches.

Twenty long years passed with only Mrs. Lotowycz's memo as an aide-mémoire about the special horticultural riches of the Laboratory grounds. Then, starting in 1995, guides touring visitors through the Lab grounds began to avail themselves of a listing of "Major Trees in Relation to Each Building at CSHL" that Frances N. Elder, a long-devoted friend of the Lab and noted conservationist, had compiled in response to popular demand. And, probably as a result of the splendid millennial exhibitions of art and sculpture that the Laboratory had sponsored here, as well as the Lab's increasingly stellar concert series, the years following 2000 saw a huge increase in the number of neighbors taking the CSHL's regularly scheduled tours. These tours now featured visits with researchers in their labs, as well as running commentaries on the Laboratory's doings past, present, and future. But by the year 2005, the Laboratory Centennial book *Houses for Science* was crying out for a successor. While *Houses* does reflect the scope of the Laboratory's scientific operation as it entered its second century back in 1990, it could not give future friends and neighbors an accurate picture of the physical plant evolving to facilitate new vistas in brain science, cancer research, and genomics in the new millennium.

In addition, site preparation had just begun for a group of six new buildings to be erected on the steep hillside high above the Lab's main thoroughfare, called Bungtown Road. CSHL and its neighboring community witnessed an abrupt change to their familiar landscape. Hundreds and hundreds of members of the "Sugar Maple Community" of common native trees were removed, both on part of the plateau at the top of the Lab's campus and on parts of the hillside leading up to it. A census had been taken before their removal – a requirement of the Environmental Impact Statement (EIS) that the Laboratory had submitted to the Incorporated Village of Laurel Hollow – and supplies of young Ash, Oak, and Maple replacements had been sourced out at many Long Island tree nurseries.

The top of the Cold Spring Harbor Laboratory, however, was bound to be looking like a moonscape for several moons to come. It seemed like the right time to start writing the sequel to *Houses for Science* that would

describe in words and photographs *both* the Lab's natural *and* built environments.

My friend and colleague at Planting Fields, Vincent Simeone, had invited me to join him and his illustrious perennial colleague Allan Armitage on a Garden Vistas excursion to some of the most wonderful gardens, public and private, of Canada's Quebec and Toronto provinces. Afterwards, Vinnie gave me permission to enroll as a non-matriculating student in his "Woodies I," the aforementioned introductory course on woody plants offered by the State University of New York at Farmingdale. Soon I could start identifying trees on my own. And exploring best buddy Faith McCurdy's garden in Cold Spring Harbor, which is filled with interesting trees that she planted herself, was confidence-building.

Achieving arboretum status was an important first step in creating the Bungtown Botanical Garden. But what does this term mean? What did we need to do?

The time was soon right to make the main campus of Cold Spring Harbor Laboratory on Bungtown Road a true arboretum. This idea of creating the "Bungtown Botanical Garden" came from studying Vinnie's Long Island University thesis detailing the creation of the Community Arboretum at the C.W. Post campus of Long Island University in nearby Greenvale. Achieving arboretum status was an important first step in creating the Bungtown Botanical Garden. But what does this term mean? What did we need to do?

"Botanical Garden . . . a garden often with greenhouses for the culture, study, and exhibition of special plants." Before I started writing *Grounds for Knowledge*, we began labeling as many trees as possible on the Laboratory's Bungtown campus. Mrs. Elder's "Major Trees" listing was greatly expanded upon by Daniel Miller, CSHL grounds manager, to include specimens of interesting woody plants, mainly trees of course, that maybe weren't so major in size but held other kinds of interest.

By spring 2007 over 150 specimens had been labeled with custom signage showing both scientific and common names. The signage effort is ongoing, but it never could have begun without the CSHL Tree Committee of Lab staffers and neighbors and professional horticulturalists.

Also, at Vinnie Simeone's suggestion the Laboratory joined the then American Association of Arboreta and Botanic Gardens (now the Public Gardens Association of America) as the Bungtown Botanical Garden. Today, in addition to having in place programs of regularly scheduled maintenance and timely replacement of trees – which has long been Danny Miller's forte – our membership responsibilities as a botanical garden include actively promoting knowledge about the trees. I hope you enjoy making their acquaintance in the pages of this introductory guide – and that you'll soon get to know them in person!

Chapters One through Five of this book are devoted to the historic Bungtown Road campus of Cold Spring Harbor Laboratory that lies within the Incorporated Village of Laurel Hollow. They will demonstrate that the development of the Laboratory's 118-acre main campus has been ever northwards along the western shore of Cold Spring Harbor, straight towards the Sand Spit, and beyond. Historically, the Laboratory had developed physically to a considerably lesser degree to the west, where it was bounded by a steep hillside of rocks and sand that the last glacier had deposited – part of the terminal moraine landscape that characterizes the North Shore of Long Island to this day. Because of this geological accident, the first five chapters flow together both chronologically and geographically. The Bungtown campus is shown in its entirety on the map in the opening pages of this book. Maps for individual chapters, which pinpoint the locations of individual trees in relation to buildings and other features of the landscape, appear early in each chapter. The new "Bungtown Botanical Garden" comprises the areas described in Chapters One through Five.

Beyond the Bungtown campus, the Laboratory has, over the past four decades, developed a number of nearby outposts – most of which lie on the east side of Cold Spring Harbor. Primarily educational and residential in character, they are the subjects of the remaining chapters of this book. Chapter Six is about the Laboratory's "think tank," the Banbury Conference Center. It was founded in 1975 on the grounds of a 45-acre Lloyd Harbor estate that was gifted to the Laboratory, together with a substantial endowment for the science at the main Bungtown campus. As shown on the Bungtown Campus Map inset, Lloyd Harbor is situated on the east side of Cold Spring Harbor's "outer harbor," where it broadens out on its way to joining the waters of Long Island Sound.

Chapter Seven describes properties later acquired directly across the harbor from the Lab in the hamlet of Cold Spring Harbor. These include firstly, at the east end of Main Street, the Dolan DNA Learning Center, which set up shop in the former Union Free School building there in 1988, adaptively reusing it for an enthusiastically received student-faculty DNA literacy program. Then, if you were to head west on Main Street, in the direction of the Lab's main campus, you would note that Main Street becomes Harbor Road as it nears the head of Cold Spring Harbor. Passing on your right the Whaler's Cove Yacht Club, which the Laboratory purchased in 1973 to prevent its rumored expansion, you would come to two handsome whaling era residences which the Lab now owns, one on each side of the road, both used now as graduate housing for students at the Laboratory's Watson School of Biological Sciences, founded in 1999.

And then, if at the traffic light immediately before the road crosses the head of harbor, you were to make a left-hand turn onto Lawrence Hill Road, shortly you would arrive at the Laboratory's Agricultural Field Station, on the south side of the road, which is the last subject of Chapter Seven, with its greenhouses, labs, and additional graduate residences.

Finally, Chapter Eight describes the Laboratory's Woodbury Genome Center, about three miles due south. With its easy access to the Long Island Expressway, the Woodbury Center is a beehive of activity. In this one industrial-strength facility the editorial and production departments of the world-renowned Cold Spring Harbor Laboratory Press share pride of place with an expanding research effort in the field of human genomics.

Supplementing the narrative text are Appendix A, an alphabetical listing of CSHL buildings with complete construction histories; Appendix B, a Bird Checklist for the Bungtown Botanical Garden, courtesy of Jim Watson; and Appendices C and D, giving clues for identifying trees by their flowers, fruits, and leaves, and suggesting the best times to see your favorite trees.

But before we begin touring the Bungtown Botanical Garden – and beyond! – here are a few more important horticultural definitions. In this guide I specify a tree's common name(s), followed by scientific name (in italics), based on "Binomial Nomenclature" – "a system of nomenclature in which each species of animal or plant receives a name of two terms, of which the first identifies the genus to which it belongs and the second the species itself." Perfected by the Swedish professor Carl Linnaeus in the mid-17th century, this system lets tree people from all corners of the globe speak to each other even to this day in what was the universal language of science in his day: Latin.

"Specimen tree" is a term usually reserved for stand-alone trees of great presence, which, back in the 18th and 19th centuries, often meant specimens of trees discovered in far away regions of China, Korea, and Japan sharing climatic similarities with the northeast seaboard region of the United States.

Among the attributes of "a noble tree," as propounded by *the* expert on woody plants, Michael Dirr, are that the tree be "immense in stature . . . architecturally elegant . . . span generations," and be inspiring, spiritual, "a skyway to heaven." The Lab is blessed with several noble trees and many striking examples of specimen trees.

Now there is just one more piece of horticultural information that is essential to know. Long Island, being surrounded by water and also subject to the warming Gulf Stream, has a distinctly different climate from the rest of New York State. In fact, it belongs to the same U.S. Department of Agriculture Hardiness Zone (Zone 7) as Cape Cod (and the Cape Islands) and the coastal regions southwest of these in Massachusetts; the southeast coast of Rhode Island; coastal New Jersey; Delaware; Maryland; much of Virginia; non-coastal North Carolina; northwestern South Carolina and northern Georgia; the list continues westward. That would put the range of average annual minimum temperatures here between 0 and 10 degrees Fahrenheit. This provides plenty of scope for the hand of man . . . and nature!

Here's a final thought. The first professional organization I joined — the American Society of Architectural Historians — took as its motto sixty years ago: "Firmitas, Utilitas, Venustas." These were the most important qualities that a building could possess, according to Roman military architect Vitruvius, writing (in *De Architectura*) early in the 1st century AD. His ideas still had currency more than fifteen hundred years later when English architectural critic John Shute (in his 1563 *First and Chief Groundes of Architecture*) called these traits "Firmness, Commoditie and Delite."

I hope to show in *Grounds for Knowledge* that, in equal measure to its buildings historic and recent, the old and new trees on the grounds of Cold Spring Harbor Laboratory possess those same qualities – of strength, usefulness, and beauty. ELW

Ballybung
Cold Spring Harbor Laboratory
Laurel Hollow, Long Island, New York
January 2008

Cold Spring Harbor L.I.

Before Bungtown, there was a Harbor . . .

The north shore of Long Island first appeared as we might now almost recognize it after the retreat of the last glacier that had covered much of the Northern Hemisphere. This happened about ten thousand years ago, or less. The melting made the seas rise to today's level, leaving behind the Long Island we know today, with its shores, plains, hills, lakes, ponds, and rivers. Particularly on the north shore of the Island there were also innumerable springs which bubbled up close to the shorelines, from water sources trapped deep underground during the course of previous glaciations. Besides these tiny streams emptying into the harbor at Cold Spring, there was a river that originated several miles south of the village. Not surprisingly, the Native American

Cold Spring Harbor, ca. 1880. Looking north from St. John's Pond. Bungtown barrel factory (red) is at far left, with Inner Harbor and Sand Spit middle distance. Inset, Hewlett-Jones grist mill (watercolor by Edward Lange; CSH Whaling Museum Collection).

inhabitants of the area called their settlement at the head of the harbor "Wawepex," which is Algonquian for "at the good little water place."

The Native Americans on Long Island's north shore had a special way of making their canoes, so essential for fishing and also trading expeditions. These were not flimsy bark-covered affairs made by bending saplings to the task, but stout watercraft that began with the fat, straight trunks of the omnipresent Tuliptree, burnt and then hollowed out. The low-lying woodlands at the edge of Cold Spring Harbor are even to this day characterized by groves of Tuliptrees and such other moisture-loving native trees as the Sweetgum, the Tupelo, and the Red, or Swamp, Maple.

"Many kinds of hickories and magnificent oaks mingled with about seventy other species of trees. This was the home of the white oak, the delight of shipwrights."

The types of trees growing higher up, in the "deciduous timber land of the rolling and glaciated northern strip," were – to quote Dr. Robert Cushman Murphy, the granddaddy of Long Island conservationists – "many kinds of hickories and magnificent oaks mingled with about seventy other species of trees. This was the home of the white oak, the delight of shipwrights."

I here quote Dr. Murphy once more, to describe what happened to the original forests after English settlers began arriving in force from New England and established most of Long Island's north shore towns before the end of the 1600s.

Old accounts, as well as old pictures, tell us that by the year 1700 the primeval woods had been mostly cleared off and that the villages, which are now shaded, then loomed up on treeless fields. We know that this bare state continued throughout the next century as well. During the Revolution the commanders of His Majesty's men-of-war, always on the lookout for provender, would post men with spyglasses in the tops, from which they could look over the bluffs and spot cattle well toward the middle of the island. . . . After the seventeenth-century clearing, much of the poor tillage was succeeded by poor stock-raising. The countryside between villages for the better part of two centuries became indifferent pasture full of half-wild cattle, horses, and sheep. . . . Now the question arises, "Where had the trees gone?"

Sailing craft north and south of Sand Spit.

In fact, trees were felled, in the beginning, not just to clear the land but to supply timber for building houses, barns, other outbuildings, fences, gristmills, churches, schools, and other needed structures. By the late 1700s shipyards were doing brisk business in Cold Spring Harbor, supplying vessels for coastwise shipping and, later, overhauling the local fleet of whaling ships. The Cold Spring Whaling Company was founded in 1834 by descendants of Thomas Jones, a British privateer who in 1695 wed Freelove Townsend, daughter of a merchant trader from Oyster Bay.

The Joneses became the first family of Cold Spring Harbor, soon becoming owners of most all the local industries – not just milling and boat building but also textile manufacture, based on wool from the local Merino sheep, and including dyeing and spinning operations housed in local factories, all dependent on water power achieved by damming the Cold Spring River early on and erecting factories at strategic points. "Tenements," that is, multiple-family dwellings, were erected on either side of the harbor to house workers in the local "Jones Industries" and their families, together with a multitude of warehouses at the harbor's edge for storing the manufactured goods.

The whaling company was ideally positioned for requisitioning locally many of the supplies needed for the eighteen-month to two-year voyages. Supplies included not only blankets, clothing, and foodstuffs for the sailors but also the materials needed for making barrels in which

The roadway over St. John's milldam (Black Walnut in ctr) was the former path of NYS Route 25A, until moved directly north across marshlands at the head of Cold Spring Harbor.

to store the whale oil processed aboard the ships during their round-the-world voyages. In fact, the barrel factory, situated about midway along the western shore of the Inner Harbor, gave the little community there its name Bungtown, after all the bungs (tapered wooden plugs) manufactured there, for stoppering barrels filled with flour and other provisions that shipped out with the whale ships. However, barrels and bungs and boats and buildings tell only part of the story of the vanished woodlands. Supplying cordwood for heating the homes of the rapidly expanding metropolis of New York depleted whatever virgin timber remained by the year 1800 and led to major depredation of the second growth as well.

Not long after the discovery of petroleum in Pennsylvania in 1859 – and the loss of several whaling vessels of the Cold Spring fleet caught in ice storms in the North Atlantic – the manufacturing base of the local economy collapsed. But all was not lost. With its picturesque scenery of hills sloping gently towards the quiet shallow waters of its protected beaches, Cold Spring Harbor, like scores of neighboring villages along the northern shore of Long Island, became "discovered" as a destination for daytime vacationers aboard excursion boats from the booming metropolis, in the years following the Civil War. Not surprisingly, North Shore vacation "cottages" – some not so very different from those in Newport, Rhode Island – were soon in vogue (on Long Island's South Shore as well), built by wealthy urbanites to take advantage of pristine waterviews. Gracious mansions were also appearing along the North Shore, and a kind of re-forestation, in the shape of extensive

To a fledgling biological teaching and research station, originally chartered by the Brooklyn Institute of Arts and Sciences, members of the local Jones family made a small grant of formerly industrial lands near the head of Cold Spring Harbor.

plantings of specimen trees, was well underway by the turn of the century, orchestrated predominately by Olmsted Brothers with contributions by other notable firms of landscape architects.

The situation in Cold Spring Harbor was unique, however, in view of its highly industrialized immediate past. Along the narrow strip of land at sea level immediately bordering the western side of the Inner Harbor, the changes were of another shape, destined to be filled not only with future horticultural but also evolutionary biological content as well. To a fledgling biological teaching and research station, originally chartered by the Brooklyn Institute of Arts and Sciences, members of the local Jones family made a small grant of formerly industrial lands near the head of Cold Spring Harbor. At the ceremony of dedication in 1904, Walter R. T. Jones shared his memories of summer vacations at Bungtown and his high hopes for its future:

> Cold Spring has experienced several distinct changes since Prime, in 1845, wrote his history of Long Island. . . . The village had long possessed two factories and a flour-mill, which were of great benefit to the neighboring farmers in taking their wool and grinding their grain; also two or three stores, all doing a small paying business. With the introduction of the whale-fishery business, the village awoke to a real boom. Buildings were erected to accommodate this business, houses built for the employees, and in my early days the village, especially on the west side, showed its activity by noises from the continued hammering of iron, the resounding echo from the coopering shops, the

clanging of boat-builders, and the buzzing of saws. When this business became no longer profitable, the place soon appeared like a deserted village – houses became vacant, buildings unused, and everywhere neglect and decay. . . .

The next change, particularly on the west side, assumed a scientific aspect. My brother, John D. Jones, inherited the family homestead and adjoining grounds. . . . The Brooklyn Institute desiring a place to establish a school of biology, he put up for that institute a building suitable for its purpose, and the school, under charge of able professors, has been a success, doing original work which has been a credit to Long Island, and acknowledged as such by similar foreign institutions. He also leased to the State of New York grounds for a fish hatchery, which is now turning out each year several hundred thousand trout and salmon to stock the inland waters of the State. Seeing the need of an organization to perpetuate the management and care of the grounds and property devoted by him to scientific research, he incorporated the Wawepex Society. . . .

Bungtown Barrel Factory with salt marsh in foreground.

> This year the Carnegie Institution, attracted by the advantages of the locality, has asked for a fifty-years' lease of part of the grounds . . . for carrying out experiments in evolution. . . . It gives great pleasure to the Wawepex Society to pass over to the representatives of the Carnegie Institution the papers putting that institution in possession of as much of the property as it desires for erecting buildings to carry out its experiments. . . . With these three institutions hailing from our village, it will assuredly soon become well known and appreciated both at home and abroad.

W. R. T. Jones' address was delivered in early June at the Jones family mansion at the head of the harbor (soon to become the Laboratory director's home) in the presence of his fellow Wawepex Society governors, Laboratory researchers, and a distinguished delegation of foreign scientists. Within two years of Mr. Jones' remarks, the Laboratory was in possession not only of an elegant new masonry Main Building for animal genetics research but also the foundations, and head house, for an extensive array of glass houses for studying plant genetics. By 1908, staff scientist Dr. George Shull was reporting in the Yearbook of the Carnegie Institution the results of his experiments crossing two different but carefully inbred strains of corn. The yield was 20 percent higher than if each strain had been allowed to open-pollinate. This was one of the first experimental demonstrations of the phenomenon of "hybrid vigor" that soon came to be employed in commercial seed production to create the high-yielding strains that today make corn the most important agricultural crop in the United States.

But we have jumped too far ahead in our story! Let's go back to that big old house still standing at the entrance to Bungtown Road, where such events were first "predicted" just after the turn of the century.

What the Laboratory is today: A research university of DNA

Eugene Blackford
Memorial Hall, 1907;
1973; 1993. Originally
dining hall and women's
dorm, later venue of first
CSH Symposium in 1933.
Inset, Davenport House,
at entrance to Bungtown.

After reaching the "Bungtown Road" directional sign on New York State Route 25A, the sight most visitors to Cold Spring Harbor Laboratory notice first is the colorful old Victorian residence at the Lab's entrance, a vision of greens and gold sitting, together with a tree of similar vintage, perilously close to the edge of the highway. After you turn into the Laboratory and leave your car or airport limo behind right there, at the Grace Auditorium parking area, be ready to partake not only of the ambience of mind-expanding science that wafts up and down the Lab's main thoroughfare, but also the sheer pleasure of touring its delightful grounds, strewn with beautiful vistas of trees and water and dozens of handsome buildings along the way. Welcome to Bungtown!

The Biological Laboratory, 1890

The Laboratory's welcoming Victorian home, the **Charles Benedict Davenport House,** was built in 1884 for the director of the Hatchery that the New York State Fisheries Commission established in Cold Spring Harbor in 1883. The new dwelling replaced an old Jones family residence that John Divine Jones (1814-1895), president of the Atlantic Mutual Insurance Company, had used as his summer house. On the back porch of Davenport House, overlooking tranquil Cold Spring Harbor, a most important piece of history occurred on a summer's

Charles Benedict Davenport House, 1884. Built for the CSH Fish Hatchery superintendent on the site of the John D. Jones house; later, Bio Lab director's residence.

day in 1889: the creation of the Biological Laboratory, one of Cold Spring Harbor Laboratory's dual antecedents. Meeting with Mr. Jones that day were Dr. Franklin Hooper, a trustee of the Brooklyn Institute of Arts and Sciences (BIAS), and Eugene Blackford, a member of the New York State Fisheries Commission as well as a BIAS trustee.

Dr. Hooper proposed that BIAS establish, on Jones family lands adjacent to those leased to the Fish Hatchery, a biological summer school like the one he had attended on Penikese Island in Buzzards Bay (off Cape Cod, Massachusetts), founded by world-renowned naturalist Louis Agassiz. Brooklyn Institute of Arts and Sciences got the green light that day from John D. Jones, and the following summer of 1890 it opened the doors of its Biological Laboratory (Bio Lab for short). In its early days it functioned as a kind of vacation school for high school biology teachers, offering courses taught by university professors who, in return, were given lab space for summer research. For labs, the Bio Lab at first utilized the fish laboratories at the Hatchery (until Mr. Jones gifted the Bio Lab with its own purpose-built laboratory in 1893; see Chapter Two). Dormitories and lecture halls were created out of abandoned factories, homes, and warehouses built decades earlier for the Jones family's enterprises in woolen manufacture and whaling.

The Laboratory's welcoming Victorian home, the Charles Benedict Davenport House, was built in 1884 for the director of the Hatchery that the New York State Fisheries Commission established in Cold Spring Harbor in 1883.

The Station for Experimental Evolution, 1904

Davenport House figures importantly again, soon after the turn of the century, when the Carnegie Institution of Washington (CIW) elected to situate its Station for Experimental Evolution immediately north of the BIAS Biological Laboratory, and John D. Jones' heirs offered his well-positioned residence to its inaugural director, Charles Benedict Davenport (1866-1944). Dr. Davenport, who had spent the preceding five summers at Cold Spring Harbor as the director of the Bio Lab, explained to the Carnegie Institution that Cold Spring Harbor possessed environmentally superior qualities ideal for plunging into a program of "experimental evolution," or scientific animal and plant breeding. Under his leadership, the Station for Experimental Evolution erected masonry labs and other facilities in rapid order along the shoreline (see Chapter Two). It wasn't until forty years later that the Carnegie Institution commissioned plans for modern facilities to be built on the uneven hillside site directly above its original buildings.

The Station for Experimental Evolution was now known (since 1920) as the Genetics Department of the CIW. Two modernistic concrete structures were built side by side for the Department, both of them completed in 1953. They were accessed from the main part of Bungtown Road, just down from Davenport House, rather than from the Lower Road next to the harbor, like the initial buildings. Massively proportioned **Milislav Demerec Laboratory** was the largest research structure yet erected at Cold Spring Harbor. It was named in honor

Milislav Demerec Laboratory, 1953; 1982; 1989. Bacterial geneticist Milislav Demerec (1895-1966) was longtime director of both the Biological Laboratory and the Carnegie Genetics Department. In foreground, White Ash.

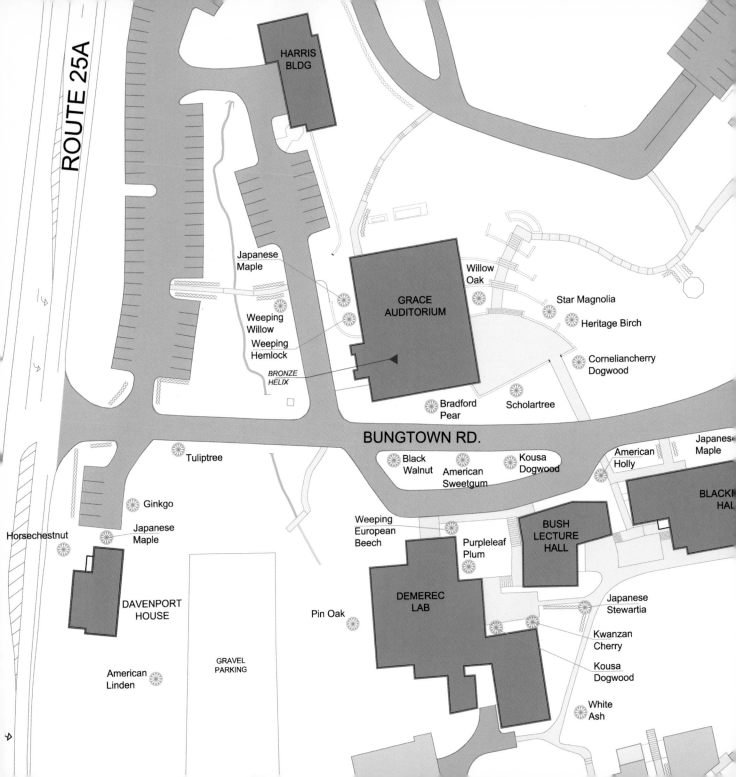

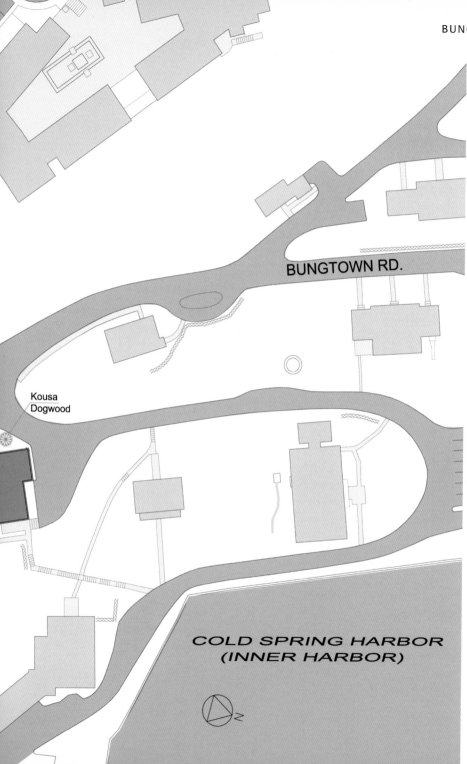

BUNGTOWN RD.

Kousa
Dogwood

COLD SPRING HARBOR
(INNER HARBOR)

N

. . . the sheer
pleasure of touring
its delightful
grounds, strewn
with beautiful
vistas of trees and
water and dozens
of handsome
buildings along
the way.

Chapter One
Bungtown Road Entrance

CSH
Cold
Spring
Harbor
Laboratory

Vannevar Bush Lecture Hall, 1953. CSH Symposia were held here from 1953 to 1986. With (l-r) Blue Atlas Cedar, Kousa Dogwood, Sweetgum.

of Milislav Demerec (1895-1966) for his long stewardship of the Carnegie Genetics Department. (Dr. Demerec directed the Bio Lab as well, during its later years.) With substantial wings added in the early and late 1980s, Demerec Laboratory became a hub of cancer related research, especially the search for cancer genes.

The second 1953 Carnegie building was the **Vannevar Bush Lecture Hall,** named in honor of National Science Foundation founding director Vannevar Bush (1890-1974). Scandinavian Modern in style, it was designed to provide a larger venue for the famous annual Cold Spring Harbor Symposia on Quantitative Biology, which had been started by the Bio Lab in the summer of 1933. During the 1953 Symposium, the first to be held in Bush Lecture Hall, James Watson presented his and Francis Crick's paper "The Structure of DNA" to an audience of virologists, polite but somewhat nonplussed by this "off-topic" information!

Cold Spring Harbor Laboratory, 1963

The Carnegie Institution of Washington withdrew from Cold Spring Harbor only ten years after the impressive new facilities were completed. To make a long story short, in 1963 it transferred its building assets to a new corporate entity called the Cold Spring Harbor Laboratory of Quantitative Biology; the last three words were later dropped. The Biological Laboratory, whose ownership in 1924 had been transferred from the Brooklyn Institute of Arts and Sciences to a local body called the Long Island Biological

Association, did likewise. The governing body for the newly minted CSHLQB (today CSHL) consisted of institutional trustees from major East Coast universities and individual trustees from the neighborhood.

Now let's get back to the Cold Spring Harbor Symposia on Quantitative Biology where we left them, in Bush Lecture Hall in 1953. If, as hinted above, this was not their original home, then what was?

Meetings and courses

Starting in 1933, long before the move to Bush Lecture Hall, the Symposia were convened for nearly twenty years in the **Eugene Blackford Memorial Hall,** which is right next door to it. Blackford Hall had been erected in 1907 in memory of the Fish Commissioner Eugene Blackford, who had helped found the Biological Laboratory. Blackford Hall's severe lines of concrete, hand mixed and hand packed into wood forms with iron bars inside for reinforcement, perhaps belies its original purpose: a Dormitory on the upper level for the women students enrolled at the Bio Lab, and a Dining Hall and Lounge for everyone on the main floor. Fireproof as well as cheap and strong, it was one of the first examples on Long Island of reinforced concrete employed for residential as opposed to utilitarian use. Except for winterization in the early 1970s, Blackford Hall remained basically unaltered until the early 1990s when substantial additions were made to the east for additional dining and to the north for enhanced kitchen facilities. The original Lounge, with its massive concrete fireplace, is hung today with pen and ink drawings that biochemist Efraim Racker, visiting in two

Blackford Hall, with White Ash in foreground.

successive summers, 1948 and 1949, made of his colleagues at Cold
Spring Harbor.

As mentioned above, the Symposia decamped for Bush
Lecture Hall in 1953 and there they remained – until it was time to
move to a larger venue again. That this later move did not
transpire until the mid-1980s was due to the role of
closed circuit television technology in permitting
expansion of the Symposia audience. Meeting sessions
could now be broadcast by CCTV to Blackford Hall
and elsewhere on the campus, for participants for
whom there were no seats in Bush Lecture Hall, with

its 150 folding chairs. In 1986 the Symposia occupied for the first time their current home, the 370 seat **Oliver and Lorraine Grace Auditorium**, which, in keeping with the environmental ethos of its day, was designed to incorporate passive cooling and heating. The main floor is partially warmed by an interior solid masonry trombe (i.e. heat-retaining) wall, which absorbs thermal energy when sunlight falls upon it from a dormer window on the main south-facing façade. This mauve-painted wall is shared by the Auditorium space and by a deep Lobby spacious enough for serving refreshments to a full house.

At its north end, the Lobby opens onto a spacious flagstone patio where meeting breaks take place in warmer weather. In the late spring, when some of the largest meetings take place, party tenting is erected over much of the patio for more chairs and video monitors. On its south side the patio is bounded by the exterior brick walls of Grace Auditorium; on the east by Bungtown Road, with a buffer of trees and shrubs in

The Octagon, 1986. Halfway point on winding walk up to Neuroscience Center.

between; and on the north and west by a steep hillside. At the extreme west end of the patio a steeply winding path takes you past **The Octagon,** with its sweeping view of the harbor below, and on upward towards various styles of visitor accommodation (and also to the Lab's original Neuroscience Center; see Chapter Four).

Due west of Grace Auditorium stands a state-of-the-art animal facility, the 1982 **Reginald Gordon Harris Building,** designed in imitation of the large old barn that stood nearby until it burned down in the early 1970s.

Reginald Gordon Harris Building, 1982. Facility resembling 19th century barn which once stood nearby. As Bio Lab director, Reginald Harris (1898-1936) founded the CSH Symposia.

A university of DNA

Getting back to Grace Auditorium for a moment, this is the place to explain that, as well as the Cold Spring Harbor Symposia, the Laboratory annually hosts over twenty additional three-to-six-day major professional meetings of 100 to 450 participants each which occur throughout the year, except for the sometimes snow-filled months of January and February. In addition, its think tank facility in neighboring Lloyd Harbor, the Banbury Conference Center, hosts over twenty meetings or workshops per year (see Chapter Six). The Laboratory also offers about thirty advanced level lecture and laboratory courses of one or three weeks' duration each, in spring, summer, and fall. Each Laboratory course is taught by a core group of instructors together with guest faculty who drop in for a day or so to give lectures, which all scientists on the grounds are welcome to attend. These training courses run concurrently during regularly scheduled sessions that fall between meeting periods. What this all adds up to is over eight thousand scientific visitors to Cold Spring Harbor Laboratory every year!

Housing on campus is limited so participants often have to be bused to large motels nearer to the center of the Island. However, all meeting and course catering arrangements take place on campus and can include al fresco dining on the lawn of Blackford Hall or a picnic supper at the Sand Spit. If, at the height of the meetings season, a cohort of Laboratory employees is factored in, then over a thousand lunches might be served of a given late spring or late summer's day out of Blackford Hall.

If, at the height of the meetings season, a cohort of Laboratory employees is factored in, then over a thousand lunches might be served of a given late spring or late summer's day out of Blackford Hall.

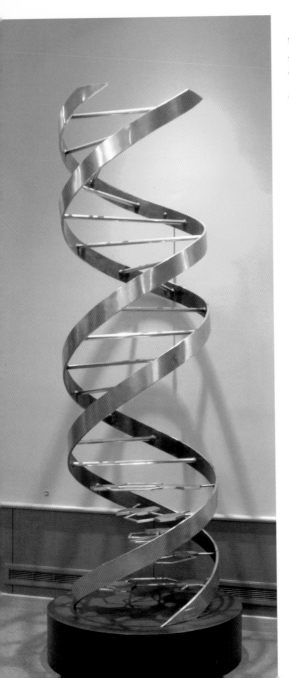

In fact, over a thousand people work full time for Cold Spring Harbor Laboratory, and of these about one third are professional scientists. They may be engaged in earning their PhDs, or employed as Postdoctoral Fellows, or serving as Principal Investigators. This last group of researchers enjoys faculty status in the Watson School of Biological Sciences, Cold Spring Harbor Laboratory's own graduate school that was founded in 1999. Traditionally, senior CSHL scientific staff also mentor those students in the graduate program of the State University of New York at Stony Brook who are doing the research for their PhDs in labs at Cold Spring Harbor.

The age of the genome arrives at CSHL

Let's go back to Grace Auditorium for one moment more. There in early June of 1993 the Laboratory unveiled the striking big ***Bronze Helix*** that enjoys pride of place in the center of the Lobby. Presented as a sixty-fifth birthday present to Dr. James Watson, it also celebrated his quarter century as director at Cold Spring Harbor Laboratory. Recently he had also launched the Human Genome Project of the

Bronze Helix, 1993; Charles Reina. Cast bronze sculpture, commissioned in celebration of the 40th anniversary of the discovery of the Double Helix and the 65th birthday of co-discoverer James Watson.

National Institutes of Health, mainly from his Bungtown Road perch but also in Congress's hallowed halls (and sometimes at a tiny desk in Bethesda!). Ten years later, almost to the day of that early June sculpture unveiling at Cold Spring Harbor, came the press conference from Washington, DC, announcing that the entire human genome had just become sequenced by the USA and its international partners.

Answers to some important questions

Q What types of research are done at CSHL?
A The majority of Cold Spring Harbor Laboratory researchers are involved in the assault on Cancer, which represents 52 percent of the annual research budget at CSHL. This is followed by Genomics and Bioinformatics, 25 percent; Neuroscience, 18 percent; and Plant Genetics, 5 percent.

Q Where do the research dollars come from?
A Besides advancing the frontiers of basic science as described above, CSHL scientists in this post-Genome age are daily making breakthroughs in genetic disease research in collaboration with the best clinical investigators worldwide. But at the same time that the possibilities for curing and preventing disease have become exponentially greater, science has become more expensive than ever. Forty years ago federal funding of science represented more than two-thirds of the research budgets of scientific institutions, with private sector donations representing less than a third. But the balance has since shifted to the exact opposite. The challenge for science funding has never been greater than it is today.

Landscape

Emblematic entrance trees

Many of the noteworthy trees in the vicinity of the Laboratory's hub flourish close to the edge of Bungtown Road. Leaf peepers beware of traffic! Let's start with those outside GRACE before venturing across the street. Out in front, two low-growing trees were planted so as not to obscure the view of this important, albeit ground-hugging, building: a **Red Cutleaf Japanese Maple**, *Acer palmatum* 'Dissectum

Copper Beech (ctr) and Japanese Maple (r) on Bungtown Road near Blackford Hall, early summer.

Atropurpureum,' with its dark red origami leaves, and a **Weeping Canadian Hemlock**, *Tsuga canadensis* 'Sargentii,' a delicate cascading evergreen. On the other side of the passenger drop-off lane, numerous examples of **Weeping Willow**, *Salix babylonica,* with their slim arching branches, hug the stream that bisects the Grace parking area. They like to "keep their feet wet" and reward us with one of the earliest signs that spring is coming when the entire grove takes on a misty yellow hue in late March.

There are five different trees worth getting to know in the vicinity of DAVENPORT across the road. Can you spot the tall, straight-trunked **Tuliptree** (also called the **Tulip Magnolia**, **Tulip Poplar**, or **Yellow Poplar**), *Liriodendron tulipifera*, a hearty Long Island native? It grows in thickets in many spots along Bungtown Road but here it is practically a specimen standing by itself, with some lower branches close enough to inspect for its distinctive leaf shape—flat on top with a slight indent in the middle. Typically the tallest tree in any landscape, the Tuliptree is adorned in late spring with large peach-colored blossoms, edged in chartreuse, that resemble the cup-shaped tulip, as well as the flowers of other members of the Magnolia family, Magnoliaceae, to which the Tuliptree belongs.

Flower of Tuliptree at Davenport House.

There is a **Ginkgo (**or **Maidenhair Tree),** *Ginkgo biloba*, standing nearby which can be identified by the way its delicate fan-shaped leaves cling to branches that point horizontally. In autumn, they slowly turn a translucent yellow and then all fall to the ground in one day, in a glowing golden puddle. It is the only surviving species of the Ginkgoacerae, a very ancient family of trees.

Growing right next to the side of DAVENPORT is a specimen of **Japanese Maple,** *Acer palmatum*, noted for its diminutive leaves so colorful in the fall. And in front is the old and stately **Common Horsechestnut,** *Aesculus hippocastanum*, which appears in most photographs of the house. Its huge leaf is palmately compound, with five fat deeply veined leaflets arranged about the stem like the fingers of a hand. These memorable looking leaves, together with the fact that they prematurely crinkle and drop due to an environmental pathogen as early as August, help identify the Horsechestnut. And if you visit during May bloom time, you'll be amazed by its tall spikes of creamy white blossoms, sitting like candles at the ends of the branches. What is more, it sloughs off big spiky brown seedballs in the fall. Even bare of leaves it is distinctive, the most "haunted" looking of trees in the winter landscape.

By the way, standing in front of Blackford, if you glance north for a moment you can't miss the huge Copper Beech, Fagus sylvatica 'Atropunicea,' overflowing the garden behind the white picket fence.

You can inspect several other typically North American trees in the middle of the grassy "Triangle" in front of DEMEREC and BUSH, including two "specimen" quality examples of Long Island natives. At the south end stands a **Black Walnut,** *Juglans nigra*, which in the fall drops big hard round chartreuse-colored nuts on your head. (Ouch! Or on your car; ping!!) It has compound leaves with 15 to 23 pointed leaflets. Its nuts are almost impossible for you or me to crack, but Dr. Barbara McClintock had in her possession a special big vise-like device of her own design for this difficult task. She knew their nutmeats were richer and tastier than the English Walnuts that you buy in the store, but more perishable. Perfect for baking Chocolate Walnut Brownies for the holidays.

Alongside the Walnut is an **American Sweetgum**, *Liquidambar styraciflua*, with its lustrous bright green star-shaped leaves and, in autumn, spiky rosy red seedballs. American Sweetgum displays brilliant shades of yellow, orange, and red fall foliage color. Incidentally, both of these native trees exhibit fuller shapes here, growing unobstructed in the Triangle, than they would if you saw them growing in the woods further down the road.

More specimen trees

A mature **Kousa Dogwood**, *Cornus kousa*, with its pretty flat pointy four-petal white flower in June, also graces the Triangle, and another grows in front of BLACKFORD, next to a Japanese Maple at the south end. In the winter these two small tree species are hard to distinguish because of their similar graceful vase-shaped habits, but in the spring they leaf out about a month apart – early April for the Maple, and early May for the Dogwood. By the way, standing in front of Blackford, if you glance north for a moment you can't miss the huge **Copper Beech**, *Fagus sylvatica* 'Atropunicea,' overflowing the garden behind the white picket fence (which belongs to Osterhout Cottage; see Chapter Three). At the north end of BLACKFORD, next to BUSH, there is a healthy planting of **American Holly**, *Ilex opaca*, to be distinguished from the English Holly found at the florists during the holidays by leaves that are distinctly less glossy and less prickly. And growing right next to the front door to DEMEREC is a dark and floppy-looking little specimen tree that is a **Weeping European Beech**, *Fagus sylvatica* 'Pendula.'

April through May is the time to wander down back into the patio between DEMEREC and BUSH where you'll experience a bevy of blooming trees, including a **Purpleleaf Plum**, *Prunus cerasifera* 'Pissardii,' a **Kwanzan Cherry**, *Prunus serrulata* 'Kwanzan' and another beautiful Kousa Dogwood. Also, just a few steps away behind BUSH is a small but special specimen, **Japanese Stewartia**, *Stewartia pseudocamellia*. Someday, when it is old enough, it will bloom in July with dainty white flowers, like miniature camellias.

On the west side of Bungtown Road, across from the Triangle, a variegated hillside planting rises from the coffee-break patio behind GRACE. Next to the building, near the foot of the steps that lead up the hill, is a grouping of **Willow Oak**,

Weeping Beech (ctr), at the front of Demerec Laboratory. Star-shaped leaves (foreground) are those of Sweeetgum.

Quercus phellos, the Oak that doesn't have deeply lobed leaves but willow shaped ones instead. You may also be able to make out the planting of **'Heritage' Birch**, *Betula nigra* 'Heritage,' diagonally across the plaza, on the lower reaches of the hillside; they have aggressively peeling light brown bark. But long before leaves appear on the Birches and the (notoriously late) Oaks, the first signs of spring

Purpleleaf Plum (top)
between Demerec
Laboratory and Bush
Lecture Hall, mid-spring.

here on the hillside are the delicate little yellow blossoms that appear on the bare-branched **Cornelian-Cherry Dogwood,** *Cornus mas*, in late March or early April. (Don't assume at first that these are Forsythia, which are later to bloom, and flourish in our woodier areas.) There are also several hillside specimens of **Star Magnolia**, *Magnolia stellata*, which show traces of bloom at about the same time but take their time before exploding into constellations of bright white.

Planted alongside GRACE and always blooming by the start of the "high season" of CSHL meetings in early May, a trio of **Bradford Pear,** *Pyrus calleryana* 'Bradford,' bear myriad snowball-like flowers. You can spot plantings of this easily recognizable tree lining parts of Cold Spring Harbor's Harbor Road and nearby Huntington's Main Street. (They also line the east side of the Beckman Laboratory parking area; see Chapter Four.) However, the Bradford Pear has a serious drawback—its co-dominant trunks, of which some may split off and lead to premature tree death. New varieties of ornamental Pear have had this tendency somewhat bred out of them, though, and would make good planting choices. There are also showy specimens of the **Scholartree** (or **Japanese Pagodatree**), *Sophora japonica [Styphnologium japonicum]*, near the entrance to the patio of GRACE, where they lend an air of enclosure. The Scholartree blossoms in mid-summer with delicate sprays of pale yellow flowers. In late summer, since it is a member of the Legume or Pea Family, Fabaceae, it is festooned with long, skinny greenish golden translucent pea

> . . . a small but special specimen, Japanese Stewartia, *Stewartia pseudocamellia*. Someday, when it is old enough, it will bloom in July with dainty white flowers, like miniature camellias.

Bradford Pear along Bungtown Road at Grace Auditorium, early spring. The Copper Beech (background) has not yet leafed out.

pods. Look for some of these still clinging to the bare branches in wintertime. (And look for more specimens of the Scholartree planted near the Gazebo; see Chapter Three.) Incidentally, both the Scholartree and the Bradford Pear have caught on as "wire-friendly" (i.e. small-in-stature) street trees all over the Island.

Scholartree at the edge of Grace Auditorium patio, late summer. Its creamy white blossoms will be transformed into skinny chartreuse "pea pods" in the fall.

Cold Spring Harbor's earliest laboratory structures

"Watson Crick (Creek)" bridge. Inset, Watson Creek near McClintock.

Now it is time to explore the oldest area of Cold Spring Harbor Laboratory, down along the shoreline near the head of the harbor. Bear right at the fork in Bungtown Road just after Blackford Hall; after you have negotiated its hairpin curve you are down at sea level on Lower Bungtown Road – right in front of Jones Laboratory.

The Bio Lab's first laboratory

A long single-story white wooden structure that looks just like a schoolhouse, the **John Divine Jones Laboratory** of the Biological Laboratory of Brooklyn Institute of Arts and Sciences was the first purpose-built laboratory structure at Cold Spring Harbor. Bio Lab co-founder John D. Jones put up the money to build it: $5,000 in 1893. He was also president of the Jones family

John Divine Jones Laboratory, 1893; 1974. John D. Jones (1814-1895) was longtime president of the Atlantic Mutual Insurance Company and Bio Lab founder. Jones Laboratory won an AIA honor award in Continued Use. Shown at low tide . . . and high.

land-holding corporation, the Wawepex Society, which as mentioned leased to the Bio Lab the land upon which Jones Laboratory was erected. Featuring wainscoted walls and a tall hipped roof punctuated with dormer windows for summertime cooling, Jones Lab originally had a large open classroom space in the center and cubicles for summer researchers along the edges.

Classes were also held in the abandoned Cold Spring Harbor Whaling Company warehouse known as the **Wawepex Building** today. (The Wawepex Society lease also included several old residences further north along Bungtown Road; see Chapter Three.) As a matter of fact, Jones Laboratory was built on the site of an especially large warehouse that first had to be demolished to make room for the new construction. This placed Jones Lab closest to the sea wall, made of stout brownstone (like the new building's foundation), which had facilitated the loading of stores such as woolen goods and the components for making barrels during the whaling era; today the CSHL dock is tethered here during the warmer months of the year. On summer evenings in the early days of the Bio Lab, the ground floor of the Wawepex Building, partially built into the hillside, was the venue for evening lantern slide shows and lectures open to the public. Today it is an office building, while the historic old laboratory next door is the venue for state-of-the-art research in neuroscience.

Wawepex Building, ca. 1835. The Wawepex Society, founded in 1895 by Jones family members, was a holding corporation for lands leased to Fish Hatchery and Bio Lab.

Earliest Carnegie buildings

It was only ten years after the construction of Jones Lab for the Brooklyn Institute's Biological Laboratory that the Carnegie Institution of Washington was considering designs for Cold Spring

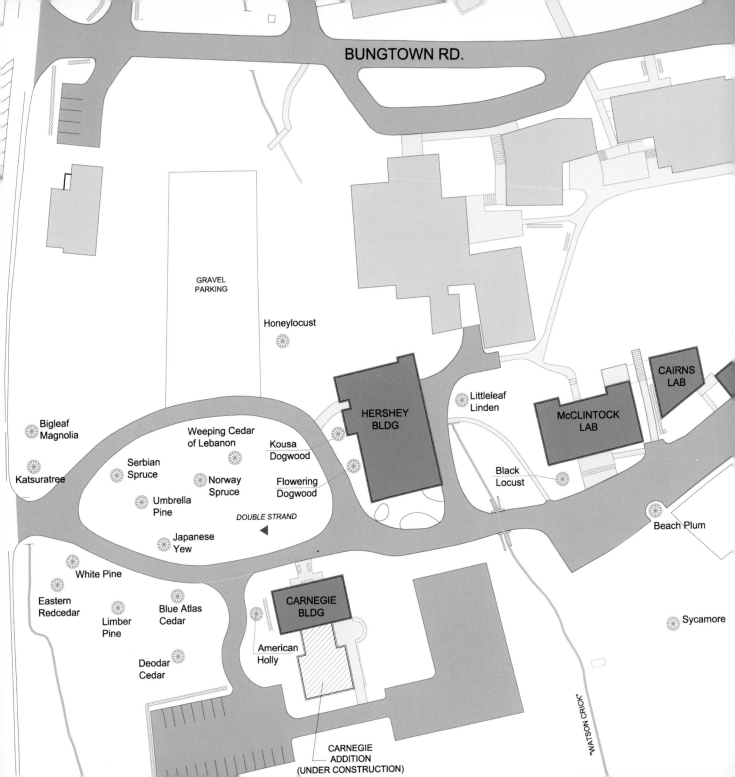

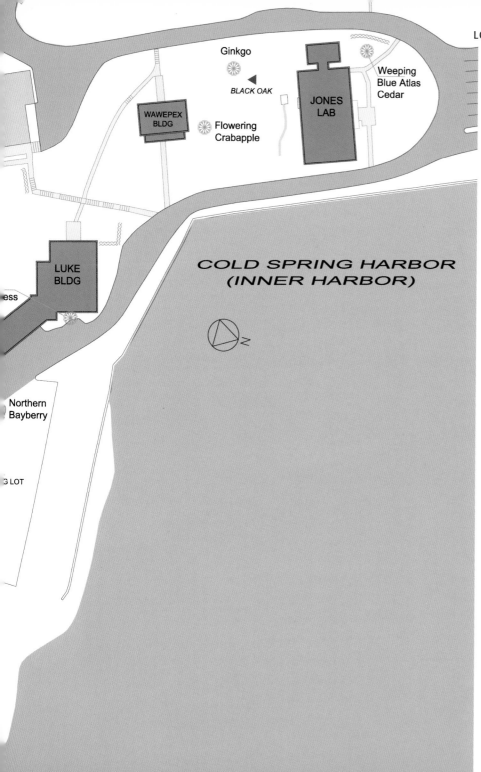

Ginkgo

BLACK OAK

WAWEPEX
BLDG

Flowering
Crabapple

JONES
LAB

Weeping
Blue Atlas
Cedar

LUKE
BLDG

ess

Northern
Bayberry

G LOT

COLD SPRING HARBOR
(INNER HARBOR)

N

A long single-story white wooden structure that looks just like a schoolhouse, the John Divine Jones Laboratory of the Biological Laboratory of Brooklyn Institute of Arts and Sciences was the first purpose-built laboratory structure at Cold Spring Harbor.

Chapter Two
Lower Bungtown Road

Cold
Spring
Harbor
Laboratory

Harbor as well, as related above (in Chapter One). To see what resulted let's march on south along Lower Bungtown Road in the direction of the head of the harbor and the lower entrance to the Cold Spring Harbor Laboratory from Route 25A. Very near the end, sitting at the eastern edge of a very broad grassy turning circle, the original laboratory of CIW pops into view. Or, alternatively, if you wanted to start from Grace Auditorium, you could look for the inconspicuous wooden staircase at the edge of Bungtown Road just south of Demerec Laboratory and take it to the bottom where it crosses the small spring-fed stream known locally as *"Watson Crick"!* (Readers will recognize this devoted reference to fellow DNA discoverers James Watson and Francis Crick.)

After restoration and construction of a new wing (completion, 2009), Carnegie Building will house CSHL archives and information services as well as new CSHL/Genentech Center for the History of Molecular Biology and Biotechnology (watercolor rendering by William H. Grover, Centerbrook Architects and Planners).

Either way, it is easy to recognize the **Carnegie Building,** completed back in 1905, because it could pass for one of the celebrated Carnegie libraries, or as a bank, with its solid-looking, classically designed stucco-with-brick-trim façade. Yet, it was designed as a laboratory and was the "Main Building" of the Station for Experimental Evolution, even housing experimental animals at first. Ever since the 1953 completion of Demerec Laboratory, however, the Carnegie Building has

Barbara McClintock Laboratory, 1914; 1987. Originally built as Carnegie Animal House, renamed in honor of Barbara McClintock (1902-1992), recipient of 1983 Nobel prize for discovering "jumping genes" in corn at CSHL. With Black Locust (r).

functioned as the main library for Cold Spring Harbor Laboratory and more recently as the headquarters of Archives and Information Services as well. In fact, the new CSHL/Genentech Center for the History of Molecular Biology and Biotechnology, including the archives of James D. Watson, Sydney Brenner, and other pioneers in the field of molecular biology, is to be housed in the wing under construction at the back of the building.

The Carnegie Building was less than ten years old when the CIW erected another stucco-and-brick structure nearby. Today called **Barbara McClintock Laboratory,** in honor of Nobel prize winning maize geneticist Barbara McClintock (1902-1992), it was built to house experimental animals, such as birds and goats, that had lived in the Main Building before. Known originally as the "Animal House" the new building featured a suite of operating rooms on the top floor. Completed in 1914, it was a stylistic copy of the world famous Zoological Station in Naples, Italy, which Dr. Davenport may have visited prior to obtaining the Carnegie Institution of Washington's commitment to Cold Spring Harbor in 1904. The Neopolitan and the Long Island buildings both feature a brick-accented stucco façade facing seawards, with two-story-tall recessed porticos in the center and three bays of

Alfred Day Hershey Building, 1906; 1979. Incorporating head house of former Carnegie Greenhouse complex. Alfred Day Hershey (1908-1997) received 1962 Nobel Prize for showing at CSHL that DNA is the molecule of heredity in viruses.

windows on either side. To preserve
its classic proportions, when in the
early 1990s a third floor was added
to McClintock Laboratory – site
today of basic genetics work – it took
the form of a recessed and mainly
glass penthouse inspired by the shape
of the monitor-roofed attic story of
the nearby Carnegie Building.

In between the Carnegie
Building and McClintock
Laboratory, and separated from the
latter by the re-emergent "Watson
Crick," stands another structure that
also dates (in part) back to the
Carnegie era: the **Alfred Day
Hershey Building,** named in honor of the CIW
researcher who proved at Cold Spring Harbor in 1952
that DNA is the molecule of heredity. The two-story
section at the back of the building, now bioinformatics
labs, was built in 1910 as the head house (potting shed)
of the CIW greenhouses; the rest of the building, completed in
1979, shares much the same footprint as the original glasshouses.
And on the north side of McClintock Laboratory is another
structure that dates back to the Carnegie era, a building that was
once a "Sheep Shed" and later the "Mouse House." With the
necessary modifications and additions, the **John Cairns
Laboratory,** as it is known today, functions as a high-powered
microscope facility.

John Cairns Laboratory,
1910; 1970. Incorporates
former Carnegie Sheep
Shed (Mouse House). John
Cairns (1922-) directed
CSHL from 1963 to 1968.

David and Fanny Luke
Building, 1913, 1950;
1999. Former Carnegie
Power House and
Carpentry Shed, united by
monumental stair hall.
Inset, Hinoki cypress at
Luke.

And immediately to the north are several more old buildings adaptively reused. The **David and Fanny Luke Building,** completed in 1999, features a new and very elegant stair hall design to unify the remodeled two-story Carnegie central steam-generating plant – the "Power House" originally – and the old single-story "Carpentry Shed" which together served for many years as CSHL's Buildings & Grounds headquarters (until the department moved in the late 1990s to the north end of Bungtown Road; see Chapter Five).

Panels of photographs illustrating successive ten year intervals in the Laboratory's now more than 110 year history encircle the central stair hall on the ground floor, which houses the Human Resources Department on one side and Public Affairs Department on the other; offices for the Department of Development are located on the second floor of the former Power House.

Landscape

The Pinetum at Bungtown

Did you know that Bungtown has a *"Pinetum,"* a place where pine trees and other coniferous evergreens can be viewed in profusion? South and east of CARNEGIE, you can view many different evergreen species, both natives and imports, all labeled; and you can study them all year-round! Let's start with the north end of the traffic circle, right in front of the building, which is dense with a naturalizing (freely self-seeding) grove of tall, deeply swoop-branched **Norway Spruce,** *Picea abies.* There is also a baby **Weeping Cedar of Lebanon,** *Cedrus libani* 'Pendula,' nearby. Somewhat bigger new plantings include **Alaska Cedar,** *Chamaecyparis nootkatensis* 'Pendula,' which is very droopy and floppy with flat blue-green foliage (and which was recently transplanted to the far side of the parking area behind the Carnegie Building to make room for new construction). There is also a young specimen of **Serbian Spruce,** *Picea omorika,* in the circle, looking like a brighter and better kempt Norway; and one of porcupine-like **Umbrella Pine,** *Sciadopitys verticillata,* with its long, thick needles radiating around each branchlet.

Plantings on the south lawn of CARNEGIE include two flourishing examples of delicately tipped, pyramidal and gracefully symmetrical Cedars: the bluish-green **Blue Atlas Cedar,** *Cedrus atlantica* 'Glauca,' and the mossy green **Deodar Cedar,** *Cedrus deodara.* There is also a **Limber Pine,** *Pinus flexilis,* with branchlets like bottle brushes. An even more unusual nearby tree is the

Bungtown "Pinetum," "a place where pine trees and other coniferous evergreens can be viewed in profusion," adjacent to the Carnegie Building.

Yellow-berried American Holly, *Ilex opaca* 'Canary.' All of these these specimens have as their backdrop groves of naturalizing **White Pine,** *Pinus strobus,* lopsided, tall and top-heavy; **Eastern Hemlock,** *Tsuga canadensis,* in the form of hedging, now grown hugely

tall; and the omnipresent **Eastern Redcedar,** *Juniperus virginiana,* which pops up all over the Lab grounds. It first appears as a prickly little shrub but can become a handsome tree as it matures, with sinewy reddish-brown bark. Finally, let's find some examples of **Japanese Yew,** *Taxus baccata.* This woody plant can become as big and tall as a Hemlock, Pine, or Redcedar, and there is an especially full one on the southeast margin of the circle that has evidently been cut back many times. You can see an example of it as a bushy shrub on the southeast margin of the Carnegie circle.

A little further away, outside the Pinetum proper, you must take a look at one last evergreen specimen, the **Hinoki Falsecypress,** *Chamaecyparis obtusa,* that was placed inside the planting berm in front of LUKE; it has a most distinctive twisted shape.

Before we leave the evergreens, however, let me give you some clues for telling them apart by comparing their needles. *Pines* have long needles in fascicles (tight bunches) of two, three, or five needles, depending on the species; White Pines have five per bunch. *Hemlocks* have flat needles and branches and very tiny cones. They resemble *Yews,* which also have flat, but darker, longer, shinier and wider needles and an outer, fleshy fruit that resembles a berry. Both Hemlocks and Yews are used extensively in hedging, but only Yews can be renewed by chopping them to the ground. *Spruces* have sharp tough fat needles in a radiating pattern. *Cedars* have fresh green fascicles of needles at their growing tips, but they are shorter and softer than dull pine needles. Incidentally, you may have noticed from its scientific name that the Eastern Redcedar is really a Juniper, not a true Cedar. This genus is much encountered at Bungtown in the form of a variety of ground-covering species that demonstrates great tenacity in clinging to steep banks.

Sculpture *Double Strand*, 1980, by Chris Solbert, in front of Carnegie Building; Bottlebrush Buckeye in foreground.

By the way, in the middle of the traffic circle part of the Arboretum I am sure you must have noticed ***Double Strand*** by sculptor Chris Solbert, who grew up in the Cold Spring Harbor area. Chris organized the Laboratory's first outdoor sculpture show, "Nothing but Steel," in 1987, of which this is one of the works original to that show. I wonder what it looks like to you?

More excellent woody plants

Now let's head over to the grassy area below and behind DAVENPORT, where there are several wonderful sights, including a lush and magnificent old **Katsuratree**, *Cercidiphyllum japonica*, and a tall and open **Bigleaf Magnolia**, *Magnolia macrophylla*. The Magnolia, growing partially shaded right next to Route 25A, is in fact clearly visible from the highway because of its foot-long leaves. But in May, most anyone would want to hike on in to examine its colossal ivory flowers at close range, from the rear garden of Davenport House. Actually from there you can spot a quartet of **Honeylocust**, *Gleditzia triacanthos*, growing in close proximity to the north side of the green aluminum extension to DEMEREC; check out the telltale thorns on all their trunks!

Blossom of Bigleaf Magnolia, at Davenport House.

Scattered about in this same general vicinity, you can also begin to look for three different species of Linden (called Lime in the United Kingdom) with their distinctive heart-shaped leaves, of sizes varying with the species. The largest leaves are

Flowering Dogwood in early spring, late spring, and fall at Hershey Building.

found on **American Linden**, or **Basswood**, *Tilia americana*, of which there is a big and beautiful old specimen several yards east of the Bigleaf Magnolia near DAVENPORT. Across the road from HERSHEY stands a nearly matched pair of **Broadleaf Linden**, *Tilia platyphyllos*. Just north of Hershey there is a **Littleleaf Linden**, *Tilia cordata*, just waiting to be noticed. And, don't miss the two Dogwoods – one Flowering the other Kousa – directly south of the building.

Now it is time to go around to the stately harbor-facing main façade of McCLINTOCK and cast an admiring glance at the very historic **Black Locust**, *Robinia pseudoacacia*, that grows in front. It can be spotted in all the old photographs of the building and of course it's now showing its age a bit. You can also see many younger Black Locusts across the road. Truly this is a species that adores the shores of Cold Spring Harbor. Now, like the Scholartree (see Chapters One and Three), Black Locust has lush green compound leaves and is a member of the Pea family and so produces "pea pods" in autumn. In wintertime, it can look a bit gnarled and spooky, like the Common Horsechestnut. Black Locust grows very straight and develops a deeply furrowed bark with ridges that appear to intersect in a diamond pane pattern. By the way, if you look in the distance, behind the younger Locusts across from McCLINTOCK, you can just make out the splotchy grey-and-tan trunk of an old **Sycamore**, or **American Planetree**, *Platanus occidentalis*, with its Maple-like leaves. However, instead of being equipped with wings like the samaras, or seed cases, of the Maples, the seeds of the Sycamore are bundles in a prickly round ball.

> You can also see many younger Black Locusts across the road. Truly this is a species that adores the shores of Cold Spring Harbor.

Behind McCLINTOCK – on the BLACKFORD lawn actually – look for the two stout examples of **White Ash**, *Fraxinus americana,* another one of Bungtown's emblematic natives. In fact, there are so many big specimens of the White Ash growing in the midst of the otherwise well-manicured southern end of Bungtown, that when all the other distinctive species have been ruled out, it is probably an Ash if most of the tree's limbs, branches, and twigs poke out at strange angles to each other. Even the trunk often grows crooked! Leaves are another clue. It could be an Ash if it is covered in big floppy compound leaves composed of five to nine tapered leaflets. At first you might confuse the White Ash with the Black Walnut, which also has big compound leaves – but more leaflets actually. However, these large leaves of the Ash grow opposite each other while those of the Walnut alternate. And, as mentioned, the Ash has an awkward if imposing shape – while the Walnut is a graceful tree, and of course its greater number of tapered leaflets are smaller and more refined than the Ash's. Incidentally, those Black Locusts all around McClintock Laboratory also have compound alternate leaves, but their leaflets are delicate little ovals. Check them all out!

In the fall, look for Beach Plum and Northern Bayberry to produce, respectively: little dark red plums for jam; and tiny ivory berries for scenting candles.

In fact, let's go back to the front of McCLINTOCK and inspect the waterfront parking area there, for it includes two kinds of species that especially like being near the ocean: **Beach Plum**, *Prunus maritima,* at the south end; and **Northern Bayberry**, *Myrica pennsylvanica,* at the north end. In the fall, look for them to produce, respectively: little dark red plums for jam; and tiny ivory berries for scenting candles. The Beach Plum, incidentally, has amazing fluffy white blooms, so in May don't be surprised if this planting looks as if it's covered in snow – an effect even more spectacular when encountered on the lonely dunes of the East End of Long Island!

Now it is time to head back up to Bungtown Road. As we round the big bend in the Lower Road it is hard to miss the reptilian-looking tree growing out of the retaining wall next to JONES. Dripping with long bluish-grey needles, it is the **Weeping Blue Atlas Cedar**, *Cedrus atlantica* 'Glauca Pendula.' Now look at the hillside behind; here is a most perfectly formed Ginkgo, plus a specimen of **Sweetbay**

Weeping Blue Atlas Cedar, Jones Laboratory.

Sculpture *Black Oak,* 2000, by Chris Solbert, near Wawepex, with Ginkgo (l) and Norway Spruce (ctr distance).

Magnolia, *Magnolia virginiana*, which has lovely big creamy flowers that bloom in early summer – instead of spring like the other Magnolias.

And standing right there also is another kind of "tree" that we will encounter when we take the higher roads at Bungtown: ***Black Oak,*** Chris Solbert's 1987 homage to the Laboratory grounds in black-painted steel!

The good works of LIBA: homes and laboratories

With a hillside to the west and a harbor to the east, the straightaway that makes up the middle section of Bungtown Road looks filled with old houses, at least to the uninitiated. Some date from the first third of the 19th century, the early Jones Industries era, and are still used for residential purposes. Other similar looking buildings are actually about a hundred years younger, having been built in the late 1920s as modest laboratories for the Biological Laboratory. This was after the Bio Lab came under the sponsorship of the Long Island Biological Association. Its trustees actively raised funds from the Laboratory's generous estate-owning neighbors to purchase additional land and erect new facilities. The new construction

W. J. V. Osterhout Cottage, ca. 1800; 1969. Reconstructed early dwelling, with Copper Beech in winter; temporary CSHL Library headquarters. Inset, Beech in summer.

73

was interspersed with old whaling era dwellings that the Wawepex Society had already put at the Bio Lab's disposal, starting with its founding in 1890.

We will begin our tour at Blackford Hall and go full steam ahead along Bungtown Road.

Federal houses

When Dr. Charles Davenport arrived at Cold Spring Harbor in the summer of 1898 he and his family took up residence in a small early 19th century dwelling, once a dormitory for male students and now called the **W. J. V. Osterhout Cottage**. Many years later the scientist Al Hershey and his family lived here before building a house in the neighborhood; they were the last to occupy the house before it was reconstructed in 1969. Originally the Lab meant to restore Osterhout Cottage, but underground springs in the area had done irreparable damage, so it

William H. Cole Cottage, 1934. Depression era building.

was rebuilt along the original lines together with a two-story addition of contemporary vintage. Afterwards Jim Watson and his growing family moved in, followed several years later by the Mike Wigler family.

Moving right along, don't neglect to observe diminutive **William H. Cole Cottage,** built in 1933, during the Depression, by otherwise unemployed laborers; it is wedged into the intersection of Bungtown Road and the Upper Road.

Another old dwelling reconstructed after water seepage problems also stands at the foot of the Upper Road, across from Cole Cottage. Named **Timothy B. Williams House** in honor of Colonel Williams, who led the Bio Lab fund drive for purchasing

Timothy Williams House, ca. 1835; 1977. Reconstructed late Federal style tenement, with steel sculptures *Midnight Fair,* 1980, and *Nuts & Bolts,* 1980, by Michael Malpass.

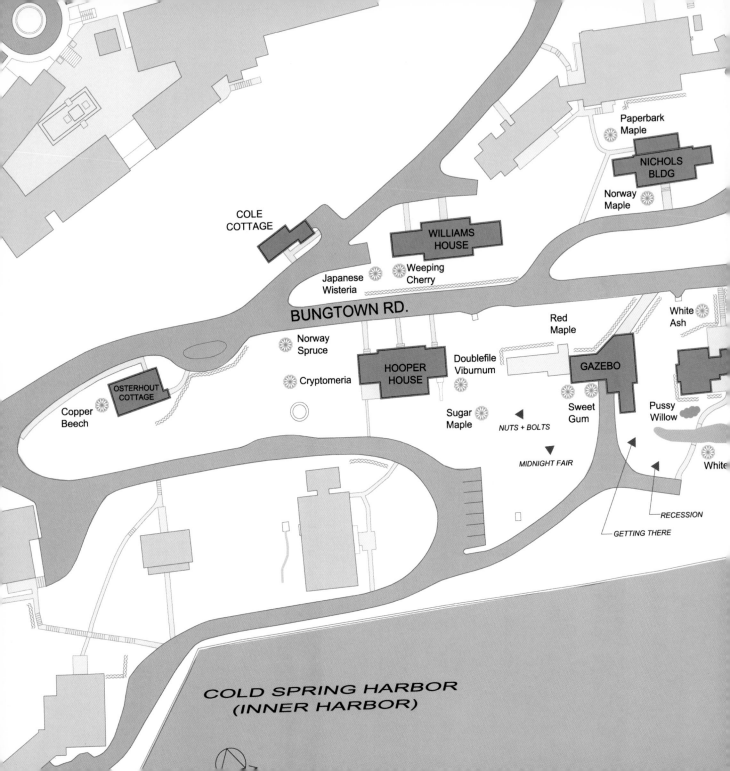

PAPERBARK
Maple

NICHOLS
BLDG

Norway
Maple

COLE
COTTAGE

WILLIAMS
HOUSE

Weeping
Cherry

Japanese
Wisteria

BUNGTOWN RD.

White
Ash

Red
Maple

Norway
Spruce

Doublefile
Viburnum

GAZEBO

Cryptomeria

HOOPER
HOUSE

OSTERHOUT
COTTAGE

Copper
Beech

Sugar
Maple

Sweet
Gum

Pussy
Willow

◀ NUTS + BOLTS

White

▼ MIDNIGHT FAIR

RECESSION

GETTING THERE

COLD SPRING HARBOR
(INNER HARBOR)

BUNGTOWN RD.

Scholartree

Incensecedar

PAGE LAB

FIRE HOUSE

Dawn Redwood

Horsechestnut

American Beech

Thornless Common Honeylocust

Black Locust

The new construction was interspersed with old whaling era dwellings that the Wawepex Society had already put at the Bio Lab's disposal, starting with its founding in 1890.

Chapter Three
Central Bungtown Road

Cold
Spring
Harbor
Laboratory

the building and its twenty-six-acre hillside site, this Colonial style dwelling originally housed textile workers in Jones Industries days. The late 1970s reconstruction created four duplex apartments within the traditional shell of the building and a dormer-windowed studio across the top floor. When needed, these centrally located accommodations can be converted to temporary office space, but for the most part Williams House still serves the residential needs of summer and temporary staff and visitors.

Across from Williams House, on the east side of Bungtown Road, look for its mate, **Franklin Hooper House;** each was designated on an old Jones family map as a "tenement," in other words, a multiple family dwelling for Bungtown Road industrial workers. Named in honor of the co-founder of the Biological Laboratory, Hooper House was used as the Laboratory's first dining hall as well as a dormitory in the early days. In the 1960s it was extensively renovated for family apartments.

Franklin Hooper House, ca. 1835, with Norway Spruce in foreground, recognizes this founder of the Bio Lab. Hooper was originally built for textile workers in the Jones industries days.

Neo-Colonial labs

Not long after the Williams House tract was acquired, a new laboratory was erected just north of it, on the same hillside but higher up. As completed in 1928, the **George Lane Nichols Memorial Building** contained an aquarium, several labs, and an animal operating room. Although it now houses central administration, CSHL's Jim Watson performed experiments in one of the labs as a graduate student in the summer of 1948; and during World War II an "Aeroliser" for administering antibiotics was developed here. The Long Island Colonial Revival style of the Nichols Building belies its original purpose but presents a cheerful face to visitors and staffers alike. Its architect, Huntington, Long Island, resident Henry H. Saylor, was well known to LIBA Chairman Arthur Page.

Saylor wrote a user-friendly *Dictionary of Architecture*, which went through many editions, and edited architecture books that were published by Doubleday, Page and Company, which Arthur Page's father had founded. Saylor designed a dormitory for the Hill School in Pennsylvania, but at heart he was a writer and editor, covering the 1920s domestic architecture scene which was dominated by two styles: bungalows, plus styles like Colonial Revival and Dutch Colonial that were modern adaptations of historical styles. So Saylor was a logical choice to interpret Chairman Page's tasteful vision of how the Bio Lab should grow, by dressing the new-fangled science in old-fashioned garb to match the Colonial buildings in the neighborhood.

George Lane Nichols Memorial Building, 1928, built as a research laboratory in Colonial Revival style, with White Ash (l and r).

Saylor's most famous design at Bungtown may well be the historic core of what is now the **Max Delbrück/Arthur W. and Walter H. Page Laboratory,** today home base to CSHL's

Max Delbrück/Arthur W. and Walter H. Page Laboratory, 1926; 1981, 1987. Max Delbrück (1906-1981), founder of the CSHL Phage Course, mentored the first generation of molecular biologists, including Jim Watson.

plant genetics research. The middle section was erected in 1926 as "Davenport Laboratory" to replace an old wooden whaling era warehouse that Bio Lab scientists had been actively using for experiments until one day it caught fire and burned to the ground.

Neo-Colonial style Davenport Laboratory is where in 1945 the legendary Phage Course taught by Max Delbrück and colleagues first began, and it was here that a brand new breed of scientist was created: the molecular biologist. The teaching lab used by the course is still in place on the second floor of this historic building, but nowadays it is flanked by two "wings" added separately in the 1980s in response to new knowledge about "jumping genes" and new technologies like recombinant DNA.

Moving experiences

To make room for the Page addition to what is now the Delbrück/Page Laboratory, a building that had already been moved once had to be moved again. Cold Spring Harbor's second **Firehouse,** originally erected in 1906 on the east side of Cold Spring Harbor, was purchased by the Bio Lab at auction for $50 in 1930. Then it was towed over to the Laboratory on a barge and deposited on a new foundation just north of the then Davenport Lab, where it became an apartment house. But there wasn't enough space between the two buildings for construction of the Page Lab addition to the north of Davenport. So in 1987 the Firehouse was once again lifted off its foundation. This time it was mounted on wooden skids so it could be pulled across a fifty-yard section of steel railroad track to its newest foundation – and a substantial interior renovation.

The addition of the south wing to what is now the Delbrück/Page Laboratory presented no such problems. Further down the road in this direction stands CSHL's **Gazebo,** a fanciful structure that rests atop what was designed as a secondary waste water (sewage) treatment facility built into the hillside in 1976; it is now decommissioned since CSHL joined

The Firehouse, 1906. Originally Cold Spring Harbor's second firehouse, it was barged across the harbor to CSHL for housing in 1930.

The Gazebo, 1976. Sits atop secondary waste-water treatment plant, now decommissioned. Won a Western Red Cedar Shingle and Shake Award. the county sewer system. From the harbor, the former waste water plant is mostly hidden by a Victorian style wood-shingled superstructure that won an award from the National Red Cedar Shingle and Shake Council. From Bungtown Road you see only the landscaped patio on top of the plant and a step-down pavilion at the back with an 180 degree panoramic overview of Cold Spring Harbor. And, the attractive finial atop the Gazebo's roof features a copper model of an adenovirus, the organism used by the Cold Spring Harbor Laboratory tumor virologists.

Landscape

Shoreline trees

Let's go down towards the water's edge and have a look at the backs of Delbrück/Page Laboratory and the Gazebo. Can you spot the hidden spring-fed pool that's behind DELBRUCK/PAGE – traversed by

Red Maple near Gazebo, with early spring red leaf buds.

the little wooden bridge? Here you can harvest watercress almost all year round. A big old **Pussy Willow,** *Salix caprea*, hugs the edge of the pool. Two lofty Sweetgums, one at each end, both loving the damp, embrace the GAZEBO – but notice the difference in their shapes. The taller, older one bears its rounded leaf bundles aloft while the shorter, younger one sports a fully pyramidal shape. (The oldster undoubtedly had plenty of competition for sunlight from scores of others like it in Bungtown; now it's a skinny survivor!)

Twirling in the wind between the Sweetgums and the heel of the Lower Road is Meryl Taradash's ***Getting There.*** A hollow alumimum "infinity" flower atop a curvy painted steel stem, it was purchased from the sculptress in the course of CSHL's millennial show "Sculpture by the Sea" in the fall of 2000. Other common natives – Black Locust, Eastern Redcedar, White Ash – dot the open area next to the water. Now have a peek through Arthur Bernhang's iron sculpture

Pussy Willow behind Delbrück/Page Lab.

Recession for a tightly framed view of just how far we've come on our walk so far.

At the northeast corner of the FIREHOUSE, close to where the marsh begins, you can inspect an example of the famous "tree [that] grows in Brooklyn," **Tree-of-Heaven,** *Ailanthus altissima*. Considered a weed by some, it is most attractive in late summer when it is covered by large red-and-yellow speckled samaras, or winged seed containers. The samaras can be spotted yards and yards away; from a distance they take on an unlikely shade of orange. This fast-growing tree has huge compound leaves made up of numerous hefty pointed leaflets, by which it can easily be identified at the edge of most highways and many other sites of disturbed soil. Look behind this big if somewhat unruly specimen to view a youngish clump of **American Beech,** *Fagus grandiflora*. Here are two ways to spot Beeches: ultra-smooth medium grey

American Sweetgum, and aluminum and steel sculpture *Getting There*, 1982, by Meryl Taradash, behind Gazebo.

Common Horsechestnut
in bloom near Firehouse.

"elephant hide" bark; and faded brown serrated oval leaves persisting through the winter, especially on young trees like these. Usually we think of Beeches as huge specimen trees with lustrous richly colored leaves (and we will encounter specimens like these in Chapter Five), but these are really baby Beeches that took root there on their own! And, right next to the Firehouse, on the south side where it gets lots of sunlight, don't miss another stately Horsechestnut.

Several unusual evergreen specimens were planted about forty years ago on the slopes you see as you continue north. You can see the high limbs of a **Dawn Redwood,** *Metasequoia glyptostroboides,* after you round the corner of the Firehouse, climb up to the edge of Bungtown Road, and crane your neck in the right direction (north). The Dawn Redwood has a perfect, if airy, pyramidal pine tree shape. If you can get close up, you'll see it is soft and silky with long needles. However, it is a deciduous tree, not an evergreen per se, since it sheds its fine needles in early winter and replaces them in late spring. The Dawn Redwood is sometimes called a "living fossil"; its common name reflects it being in existence since the Dawn of Time, although it was identified as a living species only sixty years ago in China. The Dawn Redwood doesn't mind having its feet wet, so it is no wonder the Lab's specimen next to the Firehouse has grown so tall.

The **California Incensecedar,** *Calocedrus decurrens,* on higher ground between the FIREHOUSE and DELBRUCK/PAGE, is a native of the forested mountain slopes of western North America but appears to be surviving well next to Bungtown Road. Its needles are said to be heavily scented, but the branches are too high up to test this hypothesis!

By the way, if it were late spring and you had turned south at the GAZEBO instead of north, you would not have missed the fascinating blooms of the **Doublefile Viburnum**, *Viburnum plicatum* var. *tomentosum*, at the south end of HOOPER. It blossoms in double rows of small but broad white flowers, all up and down the plant, each doublet blooming several inches above its respective leafed-out cascading branch. And on your way to this spectacle you surely would have noted the two black-painted spheres perched on the hill below the north end of the house: *Midnight Fair* (the bigger one) and *Nuts and Bolts* (aptly named!). Sculpted from found objects of steel by Michael Malpass, they were gifted to CSHL by a Malpass collector and patron at the conclusion of the "Nothing but Steel" sculpture exhibit on the Laboratory grounds in 1987.

Roadside trees

Let's walk along Bungtown Road instead of the shoreline and head south from DELBRUCK/PAGE, where we first began. In front of that Laboratory we will pass labeled specimens of the elegant Scholartree and the rough-and-ready White Ash. Across the road, an old **Norway Maple**, *Acer platanoides*, enjoys pride of place on the front lawn of NICHOLS. Many consider it a weed tree, and for sure it is a prolific "volunteer" since it offers to grow in (far too) many places; but it is a wonderful shade tree. And when all of its leaves turn a uniform and almost iridescent pale yellow in the autumn, it is truly a magnificent sight.

Now is the time to sneak around the back of the Nichols Building and try to find the **Paperbark Maple**, *Acer griseum*; no further explanation needed with such a

Weeping Higan Cherry at Williams House.

name! Let's cross Bungtown Road again and note the location of the **Red Maple,** *Acer rubrum,* straight across the street from NICHOLS and right in front of the GAZEBO terrace. Come spring, look to its delicate little crimson red flowers as a sure sign of warming, appearing at the same time as those of the Corneliancherry Dogwood and the Star Magnolia. But if it is fall, then pop into the Gazebo for a moment and have a look down at the HOOPER back lawn and at the spreading branches of the handsome specimen of **Sugar Maple** (also called **Rock Maple** or **Hard Maple**), *Acer saccharum,* a species most cherished for its multicolored leaves in the fall – yellow, orange, and red all at the same time – and whose leaf is emblazoned on the Canadian national flag. Of course the best and most delicious maple syrup comes from the Sugar Maple. Its wood is traditionally made into a kaleidoscope of gift items up New England way and is the essential building "block" for a "butcher's" cutting board.

Continuing south on Bungtown Road, we will be in front of HOOPER in a moment, and a towering Norway Spruce of the deepest green has come into view. Not only is it huge, but its lowest limbs are most unusual looking, smooth and snake-like on top, with all the needled branchlets hanging down beneath. Growing a little bit in its shadow is another remarkable evergreen, **Japanese Cryptomeria,** *Cryptomeria japonica.* An unusual specimen for this area, it has a tall but clumpy and narrow, pine-tree shape.

Even without crossing the road again, we can spy two species of woody plants on the other side of the street which are fantastic – especially when they're in full bloom. On the front lawn of WILLIAMS is a pair of beautiful **Weeping Higan Cherry,** *Prunus subhirtella* 'Pendula,' which in late April are covered in pale pink "snow." And due north, clinging to an old Black Locust stump, is an old-fashioned vine, the **Japanese Wisteria,** *Wisteria floribunda,* which in late spring can be smothered in big dangling purple blossoms. The Wisteria was one of Louis Comfort Tiffany's signature flowers in his stained glass lamps and windows (and I am sure that this one and many others along Ridge Road may have escaped from his gardens at Laurelton Hall; see Chapter Five). If you have gone in for a closer view of the Wisteria, then right here in front of Williams House you will easily spy that big Copper Beech – with or without its reddish brown leaves that fade out to brownish green during the summer – that is growing in the back garden of OSTERHOUT.

  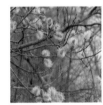

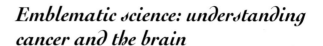

Emblematic science: understanding cancer and the brain

Sixty feet above and parallel to Bungtown Road, a narrow plateau was created over the thirty-year period 1930-60 to provide for the Bio Lab's science and housing needs. On this plateau and along the hillside leading up to it, CSHL in the 1970s-90s developed several complexes of new facilities in direct response to the "War on Cancer" declared in 1970 by President Richard Nixon, and the "Decade of the Brain" announced by President George Herbert Walker Bush in 1990. The first of these facilities to be completed clings to the hillside behind the Nichols Building.

Chestnut Oak with marble sculpture Pyramus and Thisbe, 1973, by Kenneth Campbell, near Beckman Laboratory. Inset, Zelkova in courtyard nearby.

Cancer labs on the hillside

Let's start climbing the Upper Road right where it begins at Cole Cottage. The complex we are looking for, like so many others at

Bungtown, has a historic core. In fact it incorporates the single-story 1929 **Walter B. James Memorial Laboratory,** the last building erected for the Bio Lab. A utilitarian concrete building with a flat roof, it was designed specifically for experiments studying the effect of X-rays on living cells. A good forty years elapsed before the sprawling **James Annex** was added to it in 1971 so that a newly arrived cadre of young tumor virologists could have offices and a seminar room. Many of their research labs were on the second floor, added to James Laboratory in the early 1960s, but by 1985 even more labs were needed. The research-tower-like addition to James Lab built on the hillside that

Walter B. James Memorial Laboratory (far l), 1929; 1961. Originally an x-ray facility, it became the epicenter of tumor virus research in the 1970s; with James Annex, 1971.

year was christened the **Joseph Sambrook Laboratory** in honor of Joe Sambrook, who spearheaded CSHL's tumor virus effort and wrote one of the best-selling Cold Spring Harbor Laboratory Press books of all time, *Molecular Cloning: A Laboratory Manual*. The James/Sambrook Laboratory complex remains the epicenter of CSHL's cancer research efforts.

Now that we have reached the top of the hill, we can

inspect one last building from the LIBA era. It sits directly across the James/Sambrook Circle where the Upper Road ends. **Harold Urey Cottage** has the same mid-1930s provenance as Cole Cottage at the bottom of the road, both serving to house Bio Lab scientists and visitors. Much enhanced in size and style over time, it served as the headquarters of the CSHL Press for many years and is now administrative home to the Watson School of Biological Sciences.

Joseph Sambrook Laboratory, 1985. Author of the Lab's bestselling lab manual *Molecular Cloning*, Joe Sambrook spearheaded tumor virus research at CSHL in the 1970s.

Harold Urey Cottage, 1933; 1983. Honors the discoverer of "heavy water," Bio Lab trustee Harold Urey (1893-1981). CSHL Press first set up shop here in 1983. Now headquarters for Watson School and other educational programs.

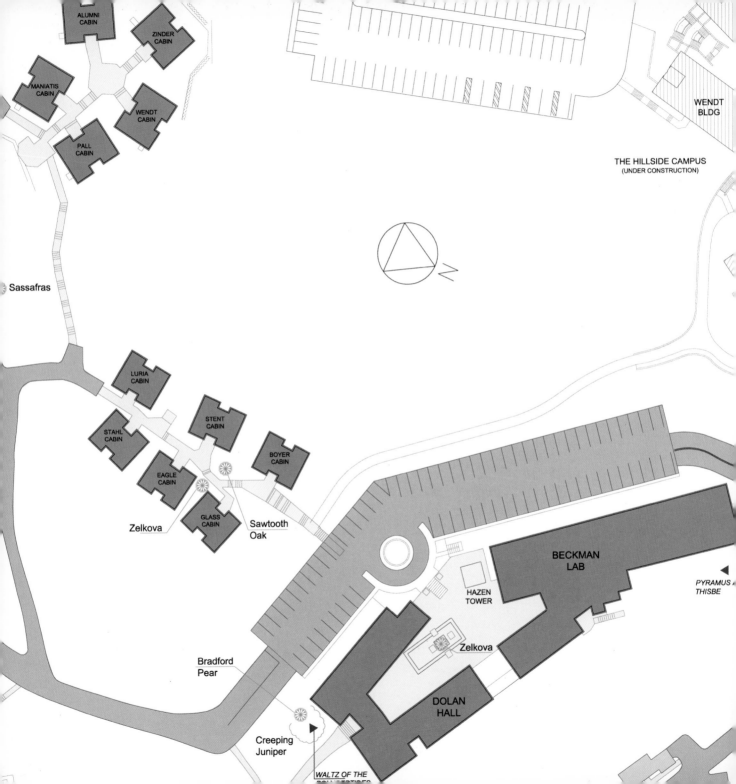

ALUMNI CABIN

ZINDER CABIN

MANIATIS CABIN

WENDT CABIN

PALL CABIN

WENDT BLDG

THE HILLSIDE CAMPUS
(UNDER CONSTRUCTION)

N

Sassafras

LURIA CABIN

STENT CABIN

STAHL CABIN

BOYER CABIN

EAGLE CABIN

Zelkova

GLASS CABIN

Sawtooth Oak

BECKMAN LAB

HAZEN TOWER

PYRAMUS & THISBE

Zelkova

Bradford Pear

DOLAN HALL

Creeping Juniper

WALTZ OF THE

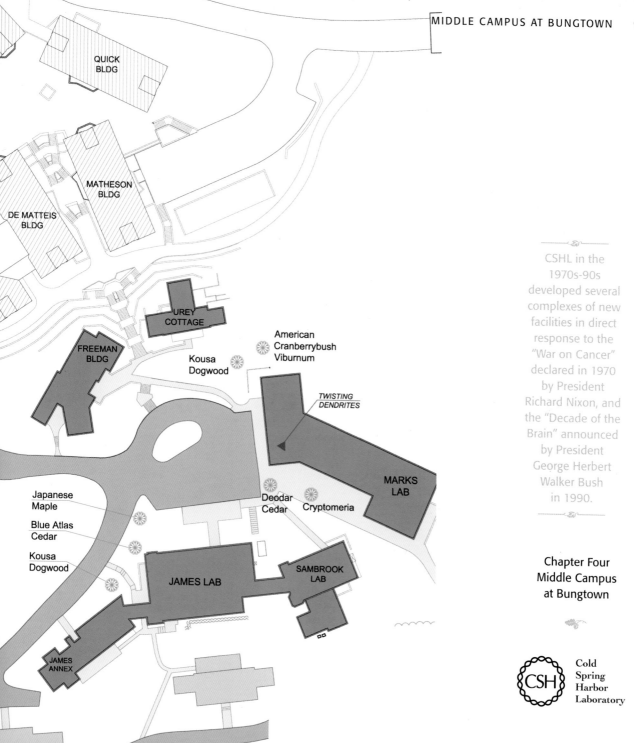

QUICK
BLDG

MATHESON
BLDG

DE MATTEIS
BLDG

UREY
COTTAGE

FREEMAN
BLDG

American
Cranberrybush
Viburnum

Kousa
Dogwood

*TWISTING
DENDRITES*

MARKS
LAB

Japanese
Maple

Deodar
Cedar

Cryptomeria

Blue Atlas
Cedar

Kousa
Dogwood

JAMES LAB

SAMBROOK
LAB

JAMES
ANNEX

CSHL in the
1970s-90s
developed several
complexes of new
facilities in direct
response to the
"War on Cancer"
declared in 1970
by President
Richard Nixon, and
the "Decade of the
Brain" announced
by President
George Herbert
Walker Bush
in 1990.

Chapter Four
Middle Campus
at Bungtown

Cold
Spring
Harbor
Laboratory

Neuroscience buildings at the top

At the same time that the Watson School was being founded, Urey Cottage was gaining neighbors on each side. To the north it is now flanked by 1999 **Edwin and Nancy Marks Laboratory** and, on the south, by 2000 **Samuel B. Freeman Building.** Marks Laboratory houses CSHL's Advanced Imaging Facility, which utilizes neuron imaging technology for research, technology development, and education. On its first floor, the

Edwin and Nancy Marks Laboratory, 1999, with Red Maple. Dale Chihuly's glass chandelier *Twisting Dendrites* hangs in this lab's skylit stair hall.

Martha Farish Gerry Room is the site of many Watson School seminars. An 800-pound "chandelier" called *Twisting Dendrites,* by renowned glass artist Dale Chihuly, sparkles in the broad skylighted stair hall. The single-story Freeman Building across the way has dormers and a large window in its gable end to shed light on the computational neuroscience and bioinformatics research going on inside its dark-brown clapboarded walls.

Due south of the Freeman/Marks/Urey complex at the top of the hill lies CSHL's original Neurobiology Center. Its two buildings were both completed in 1991 and built around a Courtyard over which the **Lita Annenberg Hazen Tower,** a square bell tower and helical staircase disguising the Neuroscience Center's chimney stack, presides. Each side of the Tower features the first letter of one of

Glass sculpture *Twisting Dendrites*, 1999, Dale Chihuly, in Marks Lab. With 800 hand-blown pieces of glass, this sculpture weighs nearly half a ton and was rigged in situ.

Chandelier-style glass sculpture symbolic of dendrites (highly branched nerve endings) as they usually appear in electron micrographs, i.e. stained a chartreuse color. Consisting of 800 hand-blown pieces of glass, weighing an average of one pound each, this work was assembled in situ by Chihuly's master riggers. First, floor-to-ceiling scaffolding was erected in the stair hall. Then the glass pieces (pre-shipped to the Lab) were individually attached to a hidden central armature of steel. Luckily the ceiling had been pre-engineered to sustain the chandelier's weight: nearly half a ton!

Samuel B. Freeman
Building, 2000, with
Zelkova.

the four bases used to make the steps in the DNA
ladder: **a**denine, **t**hymine, **g**uanine, and **c**ytosine. The
dark brick **Arnold and Mabel Beckman Laboratory,**
at the north end of the Courtyard, contains a structural biology
laboratory facility and two dedicated teaching laboratories in
addition to its neuroscience research labs. Tan brick **Charles and
Helen Dolan Hall,** at the south end, has accommodations for sixty
guests (all the rooms are graced with original – complimentary! –

Neuroscience Center, 1991, comprising Arnold and Mabel Beckman Laboratory, Lita Annenberg Hazen Tower, and Charles and Helen Dolan Hall.

watercolors by talented staff members of Centerbrook Architects and Planners of Essex, Connecticut, whose association with CSHL goes back to 1973), as well as concierge and elevator services.

At the same time that Dolan Hall was under construction, a small community of eleven log-cabin-style guest **Cabins,** each with four double bedrooms, was nearing completion in the hilly and heavily wooded area due west of Beckman Laboratory. This rustic tradition dated back to the late 1930s when a wooden "tent" community for summer visitors sprang up on the lawn behind Blackford each summer. Subsequently these were removed from that site and deposited in combos of two or three, together with plumbing connections, in the vicinity of Urey Cottage. The five resulting Cabins enjoyed pride of place where the Marks/Freeman/Urey complex is now situated, until they were demolished in the late 1980s during preparatory work for the construction of the Neurobiology Center. "Hotel-like" Dolan Hall and the other handsome buildings that make up that complex in fact now occupy the former site of late 1950s lodgings called the "Page Motel," which were removed at about the same time as the original Cabins. Besides resident scientists, visitors have always been one of CSHL's most important concerns!

Here and there are Sassafras, *Sassafras albidum,* a weedy kind of tree that can grow a mitten-shaped leaf, or three fingers, or no fingers, or all three.

The Cabins, 1989, 1991, reflect the rustic style of no longer extant 1930s cabins.

Landscape

Trees in the woods

Surrounding the CABINS in the woods above the Neurobiology Center, lush Rhododendron species flourish in the shade of native trees, mainly Oaks. Here and there are **Sassafras**, *Sassafras albidum*, a weedy kind of tree that can grow a mitten-shaped leaf, or three fingers, or no fingers, or all three. And they can be a brilliant shade of yellow or orange or red in the fall—or all three! Back at the foot of the hillside outcropping of Cabins, look for several specimens of **Sawtooth Oak**, *Quercus acutissima*. This Oak has narrow serrated leaves, instead of the "typical" amply lobed leaves, and acorns with really big bristly caps. Planted at the same time as these Oaks are a couple of exotic Zelkovas, about which more immediately below.

Plantings around the courtyard

If you would like to inspect another unusual Oak species, cross the Neuroscience Center parking area, enter the Courtyard at Hazen Tower, and exit between Beckman Laboratory and Dolan Hall. There, on the lush lawn along the back of DOLAN, stands a magnificent **Chestnut Oak,** *Quercus prinus*, right at the corner of the building. As its name hints, this Oak has chestnut-like leaves – with teeth, not deep lobes. Still, its leaves have that lustrous quality typical of many Oaks that makes them glisten when tousled by a breeze. You have probably now noticed the slender, brilliant white marble *Pyramus and Thisbe,* by Kenneth Campbell. A gift to CSHL from a generous

Japanese Zelkova in the Neuroscience Center courtyard.

North Shore couple, this sculpture of the ill-fated lovers from ancient Babylon is beautifully set off by the tall dark green Arborvitae hedge behind it.

Now, let's step back inside the Courtyard. There is one specimen here you will really enjoy: the specially sited **Japanese Zelkova,** *Zelkova serrata*. It is planted right in the center of a now picnic-table-laden "carpet" of patterned bricks, which helps anchor the tree in the broad expanse of the Courtyard. An elegant vase-shaped shade tree, Zelkova became popular some years ago for use as an American Elm substitute. And, though it can't aspire to the Elm's lofty height, the Zelkova sports brilliant scarlet fall plumage that is awe-inspiring.

Let's exit the Courtyard at DOLAN and follow the walkway over to ***The Waltz of the Polypeptides,*** artist Mara Haseltine's name for her glistening snake-like 2006 sculpture made of purple, green, gold, and blue recycled polymer. It leaps out from behind a lone Bradford Pear onto a shaggy, wiry rug of **Creeping Juniper,** *Juniperus horizontalis* 'Wiltonii.' Very hard to miss! Those purple and blue objects, each and every one of which, upon closer inspection, is clutching several golden strands, represent ribosomes in the process of assembling a protein. One of the finished products, a twisted golden mass, beckons from across the walkway. This sculpture has been evoking spontaneous comments, curiosity, and pleasure from scientists and non-scientists alike!

Sculpture The Waltz of the Polypeptides, 2006, by Mara Haseltine, outside Dolan Hall, with Bradford Pear. Iridescent recycled polymer parts portray the assembly of a protein molecule.

Specimens on the hill

Now let's go through the Courtyard one last time and exit at the BECKMAN end. First you'll see the Red Maple holding its own all by its lonesome in the planting circle. Then we will head over to MARKS to admire on its east side a big fluffy green Deodar Cedar with a most perfect pyramidal shape. Also, from here, can you spot the soaring Cryptomeria beyond the Cedar? It is stationed a bit down the hillside, near the north end of SAMBROOK. We will be going down the Upper Road in a moment, but before leaving MARKS, you should try to find the small but attractive planting of **American Cranberrybush Viburnum**, *Viburnum trilobum*, nestled into the building's southwest corner. (It is slightly hidden by a trio of Kousa Dogwood.) This plant is extra special in the autumn when it glows with big clusters of cranberry red berries and its

triple-lobed leaves turn dark red. Viburnums are fond of Bungtown, and what's not to love about this genus of woody plant? Besides the deep red fall berries and foliage, they are endowed with the sweetest smelling sparkly white spring blossoms. But before we leave the top of the hill, can you identify the new tree on the lawn in front of FREEMAN?

Viburnums are fond of Bungtown, and what's not to love about this genus of woody plant?

Near the bottom of the Upper Road, before its intersection with Bungtown Road, we'll be greeted by another Viburnum planting, here the **Leatherleaf Viburnum**, *Viburnum rhytidophyllum*, with the roughest, furriest, most deeply veined leaves imaginable. The prominent leaves of this Viburnum turn a brownish red color in late fall and they remain on the plant all winter long—an odd but comforting sight! By the way, on your way down the hill, did you notice the handsome specimens lining the road next to the JAMES ANNEX? Dating from the early part of the 1970s, these now mature plantings include a nicely broad specimen of Blue Atlas Cedar and one specimen each of the two beautifully vase-shaped small tree species, the Japanese Maple and the Kousa Dogwood.

*Maintaining the Lab and housing
its staff—and their infants*

Until the last third of the 20th century, the residential properties at the northern end of Bungtown Road were mostly under private ownership. However, two buildings had been acquired by the Biological Laboratory in the mid-1940s from heirs of Henry W. de Forest, whose vast estate lay north and west of the Sand Spit, beyond the end of Bungtown Road. This was shortly after his death in 1938 but before his magnificent estate was subdivided and the handsome mid-19th century de Forest family mansion "Nethermuir" was taken down. The first of these early acquisitions was the de Forest Stables building, which was gifted to the Bio Lab in 1942 together with the bathing beach on the Sand Spit. The second was "Airslie," the grand old farmhouse with a gambrel roof

Golden-foliage Ginkgo
with Gale House and
Ballybung, autumn.
Inset, seed pod of
American Bladdernut,
near beach road.

111

overlooking the Sand Spit, adjacent to Nethermuir, which the Bio Lab purchased the following year to be the Lab director's house. Fifty years elapsed before CSHL began purchasing various parcels adjacent to Airslie that had resulted from the postwar estate subdivision and were all improved now with residences. In the meantime, the Laboratory had acquired a number of dwellings that lay along Bungtown Road north of the Bio Lab residences and laboratories but south of the de Forest estate.

Developments south of the Sand Spit

About a quarter of a mile north of the Firehouse there is a charming old Federal "half-house" now called **Yellow House.**

Yellow House, ca. 1820, Federal style "half house."

Erected in the first third of the 19th century, it belonged to Dr. Charles Davenport's daughter, Jane Davenport Harris de Tomasi, when the Laboratory first rented it for staff housing and then eventually purchased it from Mrs. de Tomasi in 1985.

On the other side of the road, set back behind handsome old plantings, stands the **Robert H. P. Olney House,** a large, comfortable Queen Anne style residence, with the **Olney Barn,** an

elegant carriage house that inside bears the inscription "T.B.L. 1885," standing for the original builder, Timothy B. Linington. Acquired in 1973, Olney House was divided into a single-family apartment on the ground floor and a dormitory on the two top floors, while the Barn serves as base of operations for CSHL's Grounds

Robert H. P. Olney House, 1885, with Japanese Maple. Carriages once drove through its port-cochere to drop off passengers at the carriage house beyond, today the Olney Barn.

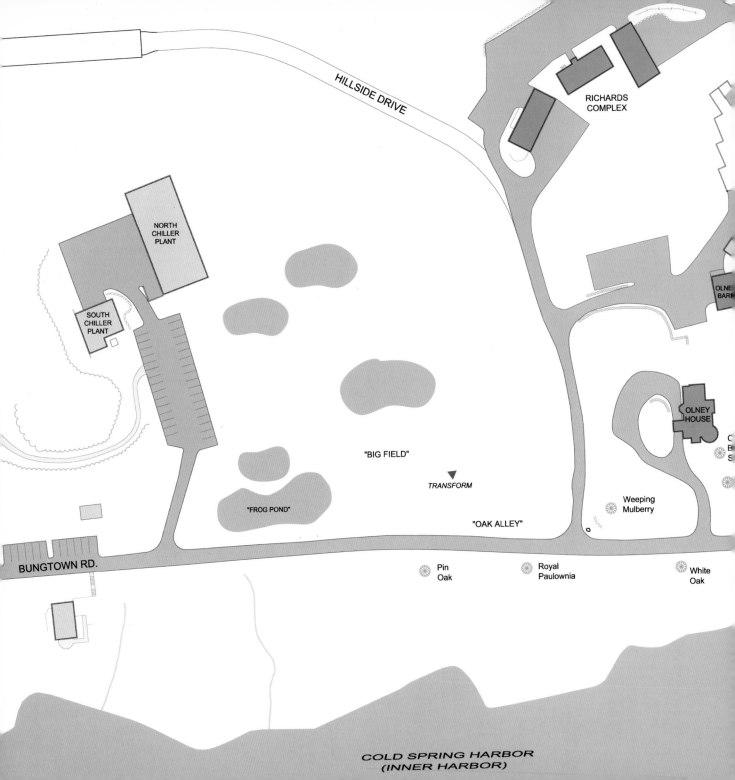

HILLSIDE DRIVE

RICHARDS
COMPLEX

NORTH
CHILLER
PLANT

SOUTH
CHILLER
PLANT

OLNE
BARI

OLNEY
HOUSE

"BIG FIELD"

▼
TRANSFORM

Weeping
Mulberry

"FROG POND"

"OAK ALLEY"

BUNGTOWN RD.

Pin
Oak

Royal
Paulownia

White
Oak

COLD SPRING HARBOR
(INNER HARBOR)

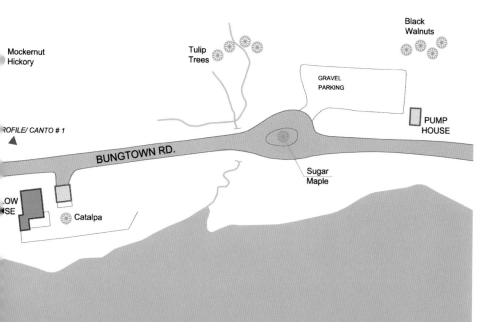

VOLLEYBALL COURTS

TENNIS COURTS

Black
Walnuts

Mockernut
Hickory

Tulip
Trees

GRAVEL
PARKING

PUMP
HOUSE

ROFILE/ CANTO # 1

BUNGTOWN RD.

Sugar
Maple

OW
SE

Catalpa

On the other side
of the road, set
back behind
handsome old
plantings, stands
the Robert H. P.
Olney House, a
large, comfortable
Queen Anne style
residence.

Chapter Five
North Bungtown Road
and de Forest Drive
(map continues overleaf)

Cold
Spring
Harbor
Laboratory

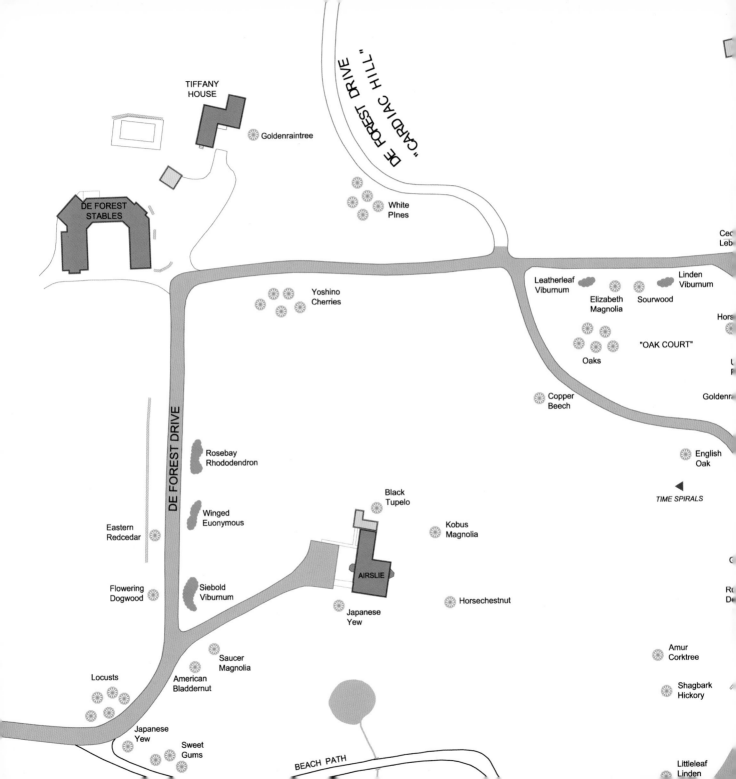

TIFFANY HOUSE

Goldenraintree

DE FOREST DRIVE "CARDIAC HILL"

DE FOREST STABLES

White PInes

Cedar Lebanon

Leatherleaf Viburnum

Linden Viburnum

Elizabeth Magnolia

Sourwood

Yoshino Cherries

"OAK COURT"

Oaks

Horse

DE FOREST DRIVE

Copper Beech

Goldenr

Rosebay Rhododendron

English Oak

Goldenra

Winged Euonymous

Black Tupelo

TIME SPIRALS

Eastern Redcedar

Kobus Magnolia

Flowering Dogwood

Siebold Viburnum

AIRSLIE

Horsechestnut

Ru De

Japanese Yew

Amur Corktree

Saucer Magnolia

Locusts

American Bladdernut

Shagbark Hickory

Japanese Yew

Sweet Gums

BEACH PATH

Littleleaf Linden

GARDEN
HOUSE

Mimosa

OLMSTED
HOUSE

Tulip Tree

Columnar
European
Beech

BALLYBUNG

Thornless
Honey
Locusts

Zelkova

DRACO

Saltspray
Roses

European
Larch

COLD SPRING HARBOR

N

The estate of Henry
Wheeler de Forest
(1856-1938)
encompassed all
the lands on either
side of what is
known today as de
Forest Drive –
formerly the main
entrance drive to
his waterfront
home.

**Chapter Five
North Bungtown Road
and de Forest Drive
(cont'd)**

Cold
Spring
Harbor
Laboratory

Department. North of the Olney Barn lies the Lab's Communal Garden, on the west end of a large restored field where, in the early 20th century days of the Station for Experimental Evolution, chickens roosted and goats grazed; on the east end are situated sand courts for beach volleyball and tennis courts for summer visitors.

Beyond the Olney Barn and nestled into the hillside to the west stands the **Jack Richards Building**, headquarters of the CSHL Facilities Department. The Richards Building, which started off life in the 1950s as a ranch style house, was remodeled after acquisition by the Laboratory in 1995 to serve the administrative needs of the Department in 1997, with barn-like structures erected at either end for Carpentry and Painting and for Mechanical Services. Further north along the road, past Olney House and the Tennis Courts, the small concrete Pump House that serviced the de Forest estate comes into view. And then the paved part of Bungtown Road ends at a set of stone pillars, and a dirt road continues on northwards to the bathing beach on the Sand Spit.

Jack Richards Building ca. 1950s; 1995. Beginning in 1971, longtime Buildings and Grounds Super-intendent Jack Richards (1925-2000) first oversaw the complete overhaul of CSHL's physical plant and then directed all new construction until retiring in 1999.

Developments north and west of the Sand Spit

The Estate Period

A paved road called de Forest Drive veers off to the west at the
pillars and leads up a gentle incline to the former Stables of the de
Forest estate, where it veers off again to the north, both turns
following the outline of the gardens around Airslie (which is
accessed immediately after the first turn). Then, after heading due
north a short distance, de Forest Drive divides in two,
one section forking off to the west and climbing the steep
hill—nicknamed "Cardiac Hill" by local joggers—to
meet Ridge Road at the top. The other section of de
Forest Drive leads in a northeasterly direction
downward towards the water and the former site of
Nethermuir. Many relics of the estate era may be found
amongst the residences sprinkled along this section of de
Forest Drive, the majority of which are now the property
of Cold Spring Harbor Laboratory.

    The estate of Henry Wheeler de Forest (1856-1938)
encompassed all the lands on either side of what is
known today as de Forest Drive – formerly the main entrance
drive to his waterfront home – and much of what were then open
fields situated at its top along Ridge Road, over a hundred acres in
all, very much to the north and west of what in those days was the
property of the Biological Laboratory. There were extensive
greenhouses at the top and staff cottages sprinkled all throughout
the estate, in addition to such typical outbuildings as the Stables
and Pump House mentioned above. From his father, Henry Grant
de Forest, Henry W. de Forest had inherited in about 1900 the

> There were
> extensive
> greenhouses at the
> top and staff
> cottages sprinkled
> all throughout the
> estate, in addition
> to such typical
> outbuildings as the
> Stables and Pump
> House.

family's Italianate style home originally built in the mid-19th century and subsequently much enlarged along appropriately Classical lines.

As related above, the de Forests called their elegant mansion **Nethermuir;** it faced northwards towards the opening of Cold Spring Harbor into Long Island Sound. **Airslie** was the name of the old farmhouse facing the Sand Spit; it was situated only about

Left: **Nethermuir, the de Forest family residence, originally built ca. 1850; demolished in 1943.**

Above, **Formal Garden adjacent to Nethermuir. (Both images, ca. 1930, from hand-colored glass plate negatives.)**

150 yards south and east of the main residence. It had been built in 1806 by Major William Jones, a cavalry officer and prominent member of the Queens County Agricultural Society. The Sand Spit that Airslie overlooks was where Major Jones exercised his prize-winning horses over two hundred years ago. In the latter part of the 19th century, Airslie was occupied by the family of Stephen Linington, brother of the Timothy Linington who, as mentioned, was the original builder of Olney House. Years later, when Airslie was the weekend home of Henry W. de Forest's sister Julia, her good friend Louise Wakeman Knox became the second wife of the famous designer in stained glass, Louis Comfort Tiffany. At Julia de Forest's invitation, they spent the first night of their November 1886 honeymoon at Airslie!

A prominent lawyer well connected to the worlds of banking, business, and the railroads, Henry W. de Forest served for fourteen years as chairman of the board of the New York Botanical Garden. So it is not surprising that he contacted Olmsted Brothers of Brookline, Massachusetts, one of the foremost landscape architectural firms of the day, for horticultural advice on his estate overlooking Cold Spring Harbor. In fact, one of the first projects he commissioned from them was a "Planting Plan for Vicinity of Miss Julia de Forest's House," in 1909. Over the course of twenty years, Olmsted Brothers generated over 350 drawings

Airslie, 1806. Federal style house originally built for gentleman farmer Major William Jones, and today residence of CSHL president.

relating to the Henry W. de Forest estate. The road system was reworked; hundreds of trees and shrubs were planted, including many exotic species new to cultivation in North America; and a formal garden area on the gentle rise to the west of Nethermuir was extensively developed.

Colleagues of Frederick Law Olmsted, Jr., in fact later considered the Formal Garden that he designed for Mr. de Forest to be one of the masterpieces of his career. It was featured in the most important gardening periodicals of the day and was the subject of hand-colored lantern slides that were disseminated through garden club circles. As these images show, the Formal Garden of Henry W. de Forest, Esq., was the quintessential spring garden, bursting with the pinks and blues and yellows of flowering trees and shrubs and massed bulb plantings. Also noteworthy in the slides are garden features such as beautifully detailed wooden arbors at the lower end, and a rustic wooden Shelter at the top; walls and archways and a free-standing Tea House of handsomely dressed stone; and beautiful hand-crafted wrought-iron gates.

Almost a hundred years later, although very few of the woody plants specified in the Olmsted plans still exist, much of the hardscape of the Formal Garden, i.e. the part of the landscape design that had to be constructed, is still extant. But first let's look at the bigger picture that includes all the de Forest estate structures that now belong to Cold Spring Harbor Laboratory.

Colleagues of Frederick Law Olmsted, Jr., in fact later considered the Formal Garden that he designed for Mr. de Forest to be one of the masterpieces of his career.

The Laboratory Era

As related in the introduction to this chapter, the Biological Laboratory acquired the **de Forest Stables,** as well as the Sand Spit, in 1942 as a gift from the Henry W. de Forest family, and the year afterwards they purchased from the de Forests what is now the residence of the president of Cold Spring Harbor Laboratory, Airslie. The de Forest Stables was a combination stable block, carriage house, and automotive garage on the ground floor and staff accommodations on the second floor which the Lab was able to put to immediate residential use. Over the years it was remodeled many times depending on the sizes of the scientists' families living on the second floor, and in 1997 the entire ground floor was renovated to become the Mary D. Lindsay Child Care Center for children and grandchildren of

De Forest Stables, 1914. Its ground floor was renovated as the Mary D. Lindsay Child Care Center in 1997.

CSHL employees. Just prior to this, in the early 1990s, the Lab had been given, as a senior staff residence, a former gardener's cottage due east of the de Forest Stables that had been sympathetically renovated in the 1950s with an addition matching the original "board and batten" style of the house. It is called **Tiffany House** after Louis Comfort Tiffany, who owned land adjacent to the de Forest estate and who built magnificent "Laurelton Hall" (destroyed by fire in 1957) after his second marriage.

Within this same time frame, Cold Spring Harbor Laboratory had obtained at auction in 1991 the parcel on which Nethermuir formerly stood, by now "improved" by a four-bedroom split-level ranch house clad variously in brick, clapboards, fieldstone, and shingles. The house had recently been seized by Federal marshals immediately upon the arrest of its then owner, an international drug trafficker

Tiffany House, ca. 1880; 1950s. Picturesque style frame dwelling.

hitherto living an almost invisible life just north of the Laboratory. Plans soon were drawn up to replace this ill-fated house with a CSHL chancellor's residence suitable for large-scale entertaining. The shape and especially the symmetry of the new house, completed in 1994 and our family's home today, were inspired by the classic farmhouses outside Venice designed in the late 16th century by the

Oaks at Ballybung, 1994. This Neo-Palladian style house, bearing a fanciful name for "place of the bung," aka Bungtown, is CSHL chancellor's residence.

Italian architect and author Andrea Palladio. Other details were adapted from small neo-classical houses outside London designed in the early 19th century by English architect and teacher Sir John Soane. Thinking back on my husband's partially Irish roots, we christened our new home at the Laboratory **"Bally-Bung,"** which we imagined just might be Gaelic for "the place of the bung," i.e. "Bungtown"!

To the east of the parcel on which Ballybung now stands, and right at the edge of the harbor, stood the former de Forest boathouse. The purchasers of the part of the de Forest estate that contained this waterside structure enlarged it in the 1950s by adding a contemporary wing comparable in size to the original building. The Laboratory acquired this property in 1996 and named it **Gale House.** And to the west of Ballybung lies the former de Forest Formal Garden that we have mentioned. When the estate was subdivided in the late 1940s, the Formal Garden area became three parcels. Today the northernmost one contains the "Tea House," which had been detailed in stone to match the garden walls, plus an attractive dwelling with natural wood shingles. This was named **Olmsted House** subsequent to its purchase in 2000 by the

Gale House, 1950s, incorporates this early 20th century boathouse from the de Forest era.

Laboratory, for joint ownership (like Tiffany House) with a senior member of the scientific staff.

Part of the rear wall of the garden is incorporated into a residence built on the middle parcel. This house also features the former rustic open-air "Shelter" at the very back of the Formal Garden. This structure was enclosed and two brick wings were added, one at each end, to create a family home back in the 1950s. It was purchased by CSHL in 2004 under a similar ownership arrangement and christened **Garden House.** Most recently, in 2007 the Laboratory purchased the southernmost Formal Garden parcel containing a wood-shingled residence, now called **Darrell House,** which incorporates at its southeast corner the octagonal sides of the stone ice house structure that had been built into the hillside and formed part of the Formal Garden wall.

Olmsted House, 1950s, dormered country-style brick dwelling a stone's throw from the former Tea House in the Formal Garden.

Top: Garden House, 1950s, incorporates the ca. 1920 wooden Garden Shelter of the de Forest Formal Garden and part of the garden wall, with new brick wings.

Bottom: Darrell House, 1950s, incorporates ca. 1920 Ice House from the de Forest estate, and retains its original octagon-shape stone detailing.

Landscape

North Bungtown roadside sightings

Our walk begins at the *"Frog Pond"* on the west side of Bungtown Road just north of the FIREHOUSE. Inspecting this former swamp puts us right at the start of *"Oak Alley,"* with many specimens of that genus flanking both sides of the road. These include labeled examples of horizontally branched **Pin Oak,** *Quercus palustris,* and an imposing old specimen of **White Oak,** *Quercus alba,* the native Oak of Long Island's North Shore. Oaks are among the last trees to leaf out in the spring and also among the last to lose their (by then rich brown) leaves in the fall. Species belonging to the White Oak group have rounded leaf lobes and sweet acorns, whereas members of the Black Oak group, especially the Pin Oak, have very deeply incised lobes and less sweet acorns. The Pin Oaks here were all planted in the mid-1970s to provide shade along Bungtown Road.

> Oaks are among the last trees to leaf out in the spring and also among the last to lose their (by then rich brown) leaves in the fall.

There are plenty of other Red Oak species here too, but it is sometimes difficult to name the exact species because members of this sub-family freely interbreed, producing many hybrids. Most of the other trees here are native or freely naturalizing. As for the White Oak sub-family, besides the White Oak species, there are also some weedy examples of Chestnut Oak, whose nuts, by the way, are especially sweet. This species generally prefers higher ground, however, like the lawn behind DOLAN in the Neuroscience Center (see Chapter Four; in fact, you will see many more if you hike up "Cardiac Hill" to Ridge Road). On the harbor side of the road, especially in the fall, you can spot the pure crimson foliage of lots of Red (Swamp) Maple, which love the moist environment there.

All along the west side of Oak Alley you can see a very large field sloping gently upwards. Thirty-five years ago, this *"Big Field"* was just a big woods, until it was opened up by cutting back the trees that had sprung up in the intervening years after the Carnegie Institution sheep no longer grazed it. There was much land-remodeling and new road construction on the upper western portion of the field in the course of the recent site preparation for the Upper Campus (see Epilogue).

The upper reaches of the Big Field have now been replanted with the Red Maple cultivar 'October Glory'; Sugar Maple (including specimens of its 'Green Mountain' cultivar); White Ash ('Autumn Purple'); and Red Oak. And in the lower section of the field closer to the road, a splendidly formed, and, in the fall, properly psychedelic Sugar Maple, with a beautifully and almost impossibly rounded crown, continues to thrive. But pride of place in the Big Field still goes to the big steel sculpture that staggers across the middle of the broad level section nearest to the road. Somewhat resembling a prehistoric Brontosaurus – or maybe a later Stonehenge? – it is Bea Perry's 1982 *Transform,* long a favorite of the denizens of Bungtown, young and old. Right across the

Sculpture *Transform*, 1982, by Beatrice Perry. Steel sculpture with rust patina, which overlooked the Hudson River before arriving at CSHL's "Big Field" in 1987.

road from it, there is a less than perfectly shaped specimen of **Royal Paulownia** or **Foxglove** (or **Empress** or **Princess**) **Tree,** *Paulownia tomentosa*, whose large purple flowers later give rise to huge seed-filled pods. If this old specimen continues in its precipitous decline, very soon it will be time to try to find another Empress (or Princess) to grace the Laboratory grounds, for its May blooms are a wondrous sight.

Now we are at the driveway into OLNEY and we'll take a quick stroll around its garden. Extra-special specimens here include lush old plantings of Japanese Maple; an extremely tall, spire-like **Fastigiate** or **Columnar Blue Spruce,** *Picea pungens* 'Glauca Fastigiata' – a National

Sculpture *Profile/Canto #1*, 1973, by Ernest Trova. Steel sculpture resting at the edge of the Olney House field.

Champion tree which in addition is mentioned in Edward Sibley Barnard's *New York City Trees: A Field Guide for the Metropolitan Area*; and a voluminously cascading **Weeping Mulberry,** *Morus alba* 'Pendula.' On the north side of the house, the unusually handsome old tree hung with a child's swing is a White Ash. In this same section of the garden look for Ernest Trova's ***Profile/Canto #1***. It arrived with the "Nothing but Steel" exhibition at the Laboratory in 1987 and has a real presence, that of a farm tractor. You can also try to make it out if you are on the other side of the road, in front of YELLOW HOUSE. Incidentally, be sure and notice the roof of that house; it is entirely covered by the branches of a large old **Common** (or **Southern** or **Eastern**) **Catalpa**, viz. **Indian Bean (or Indian Cigar) Tree,** *Catalpa bignoniodes*, a member of the legume (Fabaceae) family like the Scholartree, the Honeylocust and the Black Locust (and, noted below, the Mimosa). It grows huge heart-shaped leaves and seedpods that measure nearly a foot long by the end of the summer. And its blossoms, which appear in late June, are amazingly thick white "candles," i.e. tall columns of bloom.

Let's keep going till we come to the Bungtown Road turnabout indicated by a "Dead End" sign. It showcases a young Sugar Maple planted as a specimen tree not long ago, but it's much more fun to observe the groves of our native trees there. For example, a Tuliptree mini-forest just a few yards south of the turnabout, each tree as straight as a pencil, enjoys the marshy conditions on the west side of the road. North of the tennis parking, explore the field dotted with Dr. McClintock's fall favorite, Black Walnut. And just before Bungtown Road ends, at the stone pillars at the entrance to the beach, there are Black Locusts galore to be spotted overhead and also – to judge from the deep piles of their spikey balls at the bases of the pillars – more Sweetgum than the naked eye and the craned neck can comprehend.

De Forest Estate ramblings

West along de Forest Drive: from the beach gates up to The Stables

You can also spy the lone Walnut, Locust, or Sweetgum at the edge of the AIRSLIE lawn but there are also special legacies from the de Forest era to be viewed. Planted a very long time ago, as indicated by the Olmsted plans, are now huge-in-diameter **Japanese Yew,** *Taxus cuspidata,* at the entrances to the beach road, the Airslie driveway, and what was formerly a carriage road passing in front of the house. Also look for the imposing specimens of **Siebold Viburnum,** *Viburnum sieboldii,* which are no longer shrubby but seemingly long ago reached their small tree status – having each developed a single trunk. In late spring their delicate large white blossoms will stop you in your tracks, and by late summer, so will their rose-colored berries on rose-red stalks that eventually darken to black before dropping – or being consumed by our feathered friends!

Nearby, a very bush-like specimen of **American Bladdernut,** *Staphylea trifolia,* is another startling sight when bedecked with its freshly inflated pale-green seedpods, before they darken down with ripeness. Although it is easy to get carried away by the riches of autumn, let's think spring for a moment and imagine how very lovely it is to glance down the Airslie drive at the specimens of **Saucer Magnolia,** *Magnolia x soulangeana,*

Saucer Magnolia and Common Horsechestnut, at Airslie in April.

on either side, blooming in big beautiful cups and "saucers" of rose and white.

The north side of de Forest Drive bordering the south garden of Airslie comes alive, again in the fall, with the bright orangey-red of recent plantings of a "burning bush" cultivar of **Winged Euonymus,** *Euonymus alatus.* This woody plant takes its name from the quartet of corky "wings" running parallel to each other up and down each and every stem. Maybe you can also spot at the back of the border more shrubby reminders of the estate era in the shape of **Rosebay Rhododendron,** *Rhododendron maximum,* with its extra large leaves and extra big pale pink blossoms that bloom extra late compared to today's hybrids. By the way, you might notice, across the road on its south side, now that spring is beckoning again, the interesting rhythm between the fragile white and pink Flowering Dogwood, young and not so young, and the tough old specimens of Eastern Redcedar with their stout rusty red trunks.

North along de Forest Drive: from The Stables down to Ballybung

Just after the road turns a corner, you'll see TIFFANY and, if it's summertime, maybe you can spy the old **Goldenraintree** or **Japanese Lantern Tree,** *Koelreuteria paniculata,* bursting with pale puffy four-chambered seedpods that are even bigger than those of the Bladdernut (above), which they otherwise greatly resemble. (There is more recently planted Goldenraintree on the south side of Ballybung; see below.) In this same general vicinity of de Forest Drive, White Pine grow in profusion; there are younger ones deployed on the embankment west of the road and a number of weather-beaten old ones on the east, in the

Flowering Dogwood. Inset, Flowering Dogwood and Eastern Redcedar, en route to Mary D. Lindsay Child Care Center.

rear garden of AIRSLIE. Young apple trees are also growing in that section of the garden where there used to be an old orchard. While the apples are not yet great bloomers, a nearby planting of delicate **Yoshino Cherry,** *Prunus x yedoensis*, can be spotted all the way down at Ballybung when they are all decked out in pale pink "snowflakes."

On the east side of the road, after BALLYBUNG has come into view, perhaps you can identify the plantings of two species of Viburnum: the coarse-leaved Leatherleaf Viburnum and also **Linden Viburnum,** *Viburnum dilatatum*, which produces exceptionally attractive rich red berries towards autumn. In the midst of these tall shrubs, on the grassy verge next to the road, a recently planted **Elizabeth Magnolia,** *Magnolia x* 'Elizabeth,' is thriving—and a good thing that it is, since it was a birthday present to the author! Patented in 1977, the Elizabeth is a cross between the Cucumbertree and the Yulan Magnolia. It was the first Magnolia brought to market that was not white/pink. Its hefty, long slender buds open into generously proportioned blossoms of

Goldenraintree (Japanese Lantern Tree) at Tiffany House, summer.

Yoshino Cherry at Airslie, late spring.

creamy yellow, which happens well before the tree leafs out (as with most members of this showy genus).

The Elizabeth Magnolia grew so fast and beautifully wide that the **Sourwood,** *Oxydendron arboreum,* next to it on the bank had to be moved a little further away. Luckily it seems to have survived the move and continues to send out droopy long-lasting sprays of creamy urn-shaped blooms in the summer, and what a contrast they make with the tree's bright green summer leaves and rich red fall foliage. By the way, behind these roadside plantings, notice that the south lawn of Ballybung is populated by a number of select, stately members of the Black Oak group – I call it ***"Oak Court"*** – including a Pin Oak and some highly naturalizing Red Oaks.

Continuing on downward towards the water, let's make a quick inspection of what remains of the de Forest Formal Garden. Sadly there is virtually no woody plant material remaining there to hint at the bygone splendor of that celebrated spring-themed garden, except for disheveled remnants of what may have been the original planting of **Common Box**, *Buxus sempervirens,* that divided the lower portion into four planting quadrants. The Cherries in the wooded areas adjacent to the north garden wall might or mightn't be related to those that bloomed in the garden's glory days of the mid-1920s. But there is one tree, albeit of decidedly recent vintage, that is especially nice to look at in mid-summer, a lacy hot-pink-flowering **Mimosa** (or **Silk-tree**), *Albizia julibrissin,* growing on the north side of the Garden near OLMSTED. Chances are it is an offspring of the majestic Mimosa that used to grow next to Ballybung (in what is now the

. . . the south lawn of Ballybung is populated by a number of select, stately members of the Black Oak group – I call it "Oak Court" – including a Pin Oak and some highly naturalizing Red Oaks.

Goldenraintree's spot). It fact it is a lovely tree to visit during the entire growing season, with its unusually delicate looking leaves and, in autumn, its dainty lime green seed pods.

Mimosa at Olmsted House, mid-summer.

All around the Great Lawn: Ballybung, Gale House, and Airslie

Now we shall head to an area, at the northern end of CSHL's Bungtown campus, that has a somewhat overwhelming density of very large and notable trees, some of which undoubtedly date from the de Forest era. Be sure and note the intertwined Cedars

growing at the southern edge of the BALLYBUNG entrance
island. They share the same broad, imposing, horizontally
branched habit, but **Cedar of Lebanon,** *Cedrus libani,* is the taller of
the two, the other specimen being a Blue Atlas Mountain Cedar. In
fact, in some horticultural circles, these two Cedars, together with
the Deodar Cedar (see Chapter Two) are deemed not to be
separate species but just varieties of the single species, Atlas Cedar.
Be that as it may, both kinds of Cedars are notable for their
upright cones that are fleshy, instead of wooden, and are
pale greeny-brown.

Let's walk around the front of Ballybung, with its
matching pair of young **Columnar European Beech,**
Fagus sylvatica 'Fastigiata'; as their name indicates they
have clustered upright, instead of spreading, branches.
Around the corner, on the north side of the house, a trio
of **Thornless Honeylocust,** *Gleditsia triacanthos* var.
inermis, also of recent vintage, are already providing
delicate silhouettes and light shade on the slopes down to
the water. Near the bottom, next to the seawall, plantings
of **Saltspray** (or **Beach**) **Rose,** *Rosa rugosa,* follow the
shoreline. Their fragrance is heavenly and powerful,
indoors and out. From here you can see Kenneth
Campbell's ***Draco.*** A vertical composition of four freely sculpted
marble cubes, this sensuous piece looks like a Chinese dragon
roaring to go.

As we march on towards GALE HOUSE, can you recognize
the two specimens enroute, the Zelkova and the Goldenraintree,
near the "Hidden (Sunken) Garden"? As we are cutting through
the hedge, the handsome small tree you'll spy first is an **English**

> . . . a vertical
> composition of
> four freely sculpted
> marble cubes, this
> sensuous piece
> looks like a
> Chinese dragon
> roaring to go.

Marble sculpture *Draco*, 1965, by Kenneth Campbell, with Crabapple at Ballybung.

Oak, *Quercus robur,* with its robust little leaves, which seem to have been plucked straight from the English National Trust logo, and its deliciously perfect shape. Other newer plantings on the Gale House lawn include a vigorous 'Heritage' Birch next to the entrance and, most recently, two transplants from the vicinity of Carnegie Building. The first is a lovely Littleleaf Linden specimen. The second, marvel of marvels, is a **'Stellar Pink' Dogwood,** *Cornus x* Rutgan 'Stellar Pink,' which is a patented hybrid of the Kousa Dogwood and the Long Island native, Flowering Dogwood. Our native species, the Flowering one, was practically decimated a couple of decades back by adverse meteorological conditions termed "a blight," and that is when the planting of the Kousas, from China, Korea, and Japan, took off. The new hybrids seem to combine the best features of the two Dogwood species: the rosy pink flowers of the Flowering, and the elegant vase-shape form of the Kousa, plus the latter's superior disease resistance.

But now have a look at the two magnificent old specimen trees on the lawn closest to the house! First there is the

Time Spirals was constructed in sections near the designer's home in Scotland, and welded by Scottish fabricators who came to the USA especially to form this most perfect Double Helix. Near relatives of this sculpture may be found at the Garden of Cosmic Speculation in Dumfries, Scotland, the Museum of Life in Newcastle, England, Kew Gardens in London, and Claire College, Cambridge.

Amur Corktree, summer, with aluminum sculpture *Time Spirals,* 2000, by Charles Jencks.

colossal Ginkgo that quite possibly was planted by Henry de Forest; a seedling he may have obtained from the New York Botanical Garden could easily have been nurtured by one of the gardeners on his estate until it was ready to be planted here. I like to think it was so, as we do this often at Bungtown today. The wonderful old Ginkgo has obtained a height and girth rarely seen in our region, one for the record books.

And maybe from the same vintage is its near neighbor, the soaring **European Larch,** *Larix decidua.* Did you know that the Larch drops its needles with the approach of winter, unlike the other conifers, so that for part of the year it looks like a very dead and also very skinny tall pine tree?

HALT! Just look . . . But don't climb. It's the Lab's big glistening *Time Spirals* sculpture, about to take off. Designed in hollow burnished aluminum by international architectural critic Charles Jencks, it was commissioned as part of CSHL's millennial celebrations in 2000. Nearly twenty feet tall, it is a full-size replica of one of the DNA sculptures in Jencks' celebrated "Garden of Cosmic Speculation" in Dumfries, Scotland. His own Scottish ironmongers flew to Long Island to install it on the mound specially created for it by our Grounds Department.

In the background, behind the sculpture, try spotting the immense **Amur Corktree,** *Phellodendron amurense,* which is nearly twice as broad as it is tall. (Incidentally the Amur Corktree is totally unrelated to the cork tree that is grown commercially for bottle stoppers in Portugal.) If it happens to be summertime, discovery won't be easy because a) the leaflets of the Corktree's compound leaves are the same fresh green color as the grass; and b) it has so many full fat

Amur Corktree, summer. Inset, Amur Corktree, winter, with sculpture *Time Spirals.*

Blooming Common
Horsechestnut, at Airslie,
spring. Copper Beech,
newly leafed out, at left.

Common Horsechestnut,
Black Tupelo (l), and
Rosebay Rhododendron
(foreground) at Airslie,
fall.

branches sweeping the ground that it is very hard to spot its massive "corky" trunk. This may be the biggest specimen of its kind on Long Island and probably has a provenance similar to that of the nearby Ginkgo. What is more, in fall the leaves of both trees turn such stunning shades of pale gold that they light up the whole landscape, if only for a few precious days. And long after Mr. de Forest added it to his personal arboretum on the shore of Cold Spring Harbor, the Amur Corktree, again like the Ginkgo, was singled out for street-tree potential. I have begun noticing specimens of it along the streets of Manhattan, so it must be very tough and probably doesn't reach such tremendous proportions under adverse conditions!

So the accuracy of a tree's name can depend on whether the season is spring or fall, not just on whether or not it's decked out in its flowers/fruits/balloons/other decorations.

Let's now go and explore the north lawn of AIRSLIE. But first I would like to take you on a slight detour towards a small grove of very big Littleleaf Linden. These trees, with their thick growth of heart-shaped leaves and neatly rounded tops, provide the excellent shade adjacent to the Lab's picnic area near to the beach. So if we head towards the "woods" behind the Amur Corktree we'll be going in the right direction. In fact, directly behind the Corktree and front and center of the Lindens is one of our more unusual native trees: a **Shagbark Hickory,** *Carya ovata,* probably the largest one you can see at the Lab. Just look at that stout shaggy trunk, peeling off in really big, long strips!

It's time to head to the gigantic Common Horsechestnut on the Airslie lawn. It has several big branches that have curved down to the ground and then grown vertically back up to quite a height. The tree lost one of these large branches during a hurricane in the

mid-1970s and the resultant gap in the lower silhouette of the tree is still there (though slowly being breached by new growth). Do you see it? By the way, there is another huge tree right near the Horsechestnut but it is much harder to make out because the land behind it doesn't slope down to a beach and it is so very large you can't conceive it to be one single whole tree. It is our famous **Kobus Magnolia**, *Magnolia kobus*, originally from Japan. According to the record books, it must be the largest of its species on Long Island. And no wonder, since it is clearly indicated on an Olmsted plan dating back to 1906. Incidentally, this tree is splendidly easy to notice in early spring. It looks like an

Kobus Magnolia at Airslie, early spring.

immense cotton ball – it is so entirely covered in petite white petaled flowers – before almost any other of our deciduous woody plants have shown any signs whatsoever of coming back to life.

Now, to go from the truly sublime to the fairly ridiculous in terms of tree nomenclature, we will take a shortcut back to BALLYBUNG by heading in the direction of the apple orchard behind Airslie. As we pass the trellised area at the back of the house, we'll practically brush up against a tallish narrowly pyramidal specimen of **Black Tupelo** (or, depending on which state you hail from, **Black Gum**, **Sour Gum**, **Pepperidge**, or **Beetlebung**), *Nyssa sylvatica*, another one of those moisture loving natives. In the fall its glossy leaves usually turn an impossibly uniform shade of red, hopefully bright rather than dull, which of course depends entirely on the most recent meteorological happenings at Bungtown.

Let's now get back on de Forest Drive and head down once again towards the water. Be sure to stop and look for the majestic Copper Beech at the side of driveway to GALE, yet another estate era tree. Like its considerably younger relative way back at Osterhout Cottage, this specimen looks blackish-red when it first leafs out in the spring, and then turns to purplish-green. By late summer, it is often just plain green. So the accuracy of a tree's name can depend on whether the season is spring or fall, not just on whether or not it's decked out in its flowers/fruits/balloons/other decorations.

Now we are back near the landscaped island at the entrance to BALLYBUNG where the two beautiful old Cedars enjoy pride of place. But in actual fact, they are dwarfed in height by the two trees facing the house whose stalwart trunks are the bearers high up of flood lights for when it is dark or stormy outside. The first of these trees is an old favorite, the Common Horsechestnut, like the one in front of Davenport House at the entrance to Cold Spring Harbor Laboratory. The second is one of a side-by-side Tuliptree pair which, if I had to choose, I would say is my favorite tree, for three reasons: its absolutely ramrod straight smooth trunk; its gratifyingly easy-to-recognize truncated leaves; and its dazzlingly bright flowers, so tantalizingly beyond reach. And if, after a spring rainstorm, you could stand in our driveway cradling one of these blossoms in your hand, you surely would feel the magic of its limpid peach-like color, a reflection in miniature of our home . . . the "orange-colored house" at the end of Bungtown Road.

The Laboratory's "Think Tank"

Cold Spring Harbor Laboratory's main Bungtown campus, which we have been touring, is situated on the western shore of Cold Spring Harbor, in the Incorporated Village of Laurel Hollow in Nassau County. But in the early 1970s the Laboratory acquired a beautiful estate across the water on the harbor's eastern shore, belonging to the Incorporated Village of Lloyd Harbor in Suffolk County. The estate's creator and owner, Charles Sammis Robertson, gifted the Laboratory with his own home and several other structures in close proximity to it. By preserving the main residence and adaptively reusing the other buildings, the Laboratory created a small "think tank"-style conference center. They named the center after Banbury Lane, the graceful, long tree-lined entrance drive to the original estate

Charles Sammis Robertson House, 1936. Elegant Georgian Revival style accommodation for participants in Banbury Center meetings. Inset, Meier House, ca. 1970. Neo Georgian Revival residence, now writing center for CSHL authors.

153

off West Neck Road, which dated back to colonial times. Following is the story of how the gift came about and how the words "Banbury meeting" and "Banbury workshop" have come to underscore Cold Spring Harbor's international reputation for state-of-the-art science.

The Robertson Estate

Charles Sammis Robertson, a Long Island native and descendant of the well-known Sammis family, had assembled in the 1930s a fifty-acre estate comprising many of his family's ancestral farmlands, in a wide swath that extended due west along Banbury Lane all the way to the eastern shoreline of Cold Spring Harbor. He commissioned in

Back of Robertson House overlooks Cold Spring Harbor.

the mid-1930s a beautiful Georgian Revival home from residential architect Mott B. Schmitt; celebrated for his country houses, Schmidt also designed the Wagner wing of Gracie Mansion (official home of the mayor of New York City). The **Charles Sammis Robertson House,** clad in white-washed brick and featuring two matching wings to either side, affords an almost panoramic view of the harbor from its flagstone patio.

Upon the premature death of his wife, Marie Hoffman Robertson, who was an heir to the Great Atlantic and Pacific ("A & P") Tea Company, Mr. Robertson actively sought out an institution which could preserve and make use of his beloved homestead, where he and Mrs. Robertson had raised their own family. In 1972, after discussion with a close friend, Mr. Robertson chose to make a sizeable gift to the research endowment of Cold Spring Harbor Laboratory and the following year donated his estate to the Laboratory, fully endowing Robertson House for use related to a future conference center on the site.

Opened in 1977 by Nobelist Francis Crick, the Banbury Center would hold small meetings, with participation by invitation only, that would concentrate on cutting edge topics in biology.

The Meeting Center

The second largest building on the estate was an elegant seven-car garage built, like the residence, along classically symmetrical lines. Nestled among trees, it is situated at the far end of the immense lawn stretching to the east of Robertson House. The Laboratory immediately commissioned plans for converting the garage into the Banbury **Meeting House,** complete with a spacious seminar room in the center three-bay section plus administrative offices and a library in the flanking two-bay wings. Opened in 1977 by Nobelist

Francis Crick, the Banbury Center would hold
small meetings, with participation by invitation
only, that would concentrate on cutting edge
topics in biology. Restricting
participation to thirty-six top class
scientists ensured that each topic was subjected to
critical review and in-depth discussions. Early topics
focused on carcinogenesis and environmental hazards,

The Meeting House,
former seven-car garage
of the Robertson estate,
1936. Renovated for
adaptive reuse as a
meeting facility in 1977.
Inset, Participants on
Meeting House patio.

but later the program was expanded to cover a wide range of topics in molecular biology, human genetics, and neuroscience. These meetings established the Center's worldwide reputation as the "think tank" of modern biology. In addition, the Center held a highly influential series of meetings for lawmakers and journalists on molecular biology topics (often controversial) of importance to the public. By its thirtieth anniversary in 2007, over 500 meetings had been held at Banbury.

While thirty-six participants could comfortably congregate in the Meeting House, only sixteen could sleep in Robertson House. The Conference Center needed more bedrooms. To solve this problem noted architect Charles W. Moore designed elegantly symmetrical **Sammis Hall,** a stuccoed dormitory residence that is a Post-Modern copy of Palladio's Villa Poiana in Vicenza, Italy. It is situated behind an old apple orchard, about halfway between Robertson House and the Meeting House.

Also included in the Robertson gift was the **Superintendent's Cottage,** a charming wooden building with an inviting front porch. It is used for senior staff housing, as is the pretty **Rose Cottage,** originally built for Mr. Robertson's sister, at the end of the Big Beech Field. A new Georgian brick residence, **Meier House,** that Mr. Robertson's youngest daughter built in

Sammis Hall, 1981. Post Modern adaptation of Palladio's 1549 Villa Poiana in Vicenza, Italy.

Entrance to Banbury Center. Superintendent's Cottage, 1936, is beyond.

the 1970s, was, after purchase by the Laboratory, recently adapted for use as a writing center where authors of textbooks commissioned by the Cold Spring Harbor Laboratory Press can come together for productive long weekends in the peace and quiet of the Banbury estate.

Landscape

Signature plants

The signature plant of the entrance drive to Banbury Center is the stately Beech; plantings of American Beech, European Beech, and Copper Beech are intermingled in the fields that line the drive.

Copper Beech at Banbury Center.

Here and there you can find specimens of lacy-looking **Fernleaf Beech,** *Fagus sylvatica* 'Laciniata.' Towering Norway Spruce guard the edges of the front lawn of Robertson House while white and pink Flowering Dogwood and Rosebay Rhododendron flourish in the shady environment underneath. Other lovely elements of the Robertson estate landscape include stately Hickories and Oaks, both native species. In fact, two tall slim **Pignut Hickory,** *Carya glabra*, frame the formal entrance to the house, their leaves superficially resembling those of the Horsechestnut, but with more of a separation between the five fat leaflets. But the most spectacular natives at Banbury can be seen if you hike on down the road behind Robertson House that takes you to the beach and pool. The parking area there is nestled in a forest primeval of towering Tuliptrees.

However, the list of beautiful and unusual specimens at Banbury is so great – including American Elm, Franklinia, Oriental Spruce, and Silver Maple – that it almost needs a tree labeling program of its own for friends and neighbors!

Classrooms – and living rooms – for tomorrow's scientists

THE DOLAN DNA LEARNING CENTER ON MAIN STREET, COLD SPRING HARBOR

The educational outreach of Cold Spring Harbor Laboratory is not limited to the courses and meetings which graduate students and PhDs alike flock to ten months out of the year. Nor, to the think-tank type meetings at the Banbury Center to which only select professionals are invited. Starting in 1985, the Laboratory has become increasingly involved in DNA education at the high school, middle school, and elementary school levels through its sponsorship of the DNA Learning Center on Main Street, Cold Spring Harbor. Founded by David A. Micklos, a staffer from the Laboratory's Public Information department who came with both a public relations and

Dolan DNA Learning Center, 1925, 1988, 2001. Former Union Free School of Cold Spring Harbor, renovated and enlarged for DNA literacy program classrooms and labs; street facade.

teaching background, the earliest days of the DNA Literacy Program saw mobile "Vector Van" laboratories reaching out to high school biology teachers in need of curricula introducing the exciting new world of DNA science to their students. Shortly thereafter the program was able to move into permanent headquarters in a de-accessioned school building in the hamlet of Cold Spring Harbor, close to the spot where Goose Hill Road

enters Main Street. The former Union Free School was soon equipped with several state-of-the-art labs where morning and afternoon programs run continuously throughout the school year, and nowadays need to be booked months in advance.

The demand for DNA literacy grows ever faster. Teacher training workshops have extended the Laboratory's reach to forty-two states and into Great Britain, Italy, Sweden, Russia, Singapore, and Australia. And on Main Street in Cold Spring Harbor, an addition of such major proportions was added in 2001 that this birthplace of globe-girdling DNA education was renamed the **Dolan DNA Learning Center.** Now, in addition to hands-on experiments and multimedia presentations in the retro-fitted former school auditorium, the lab field trips include interactive computer experiences and ground-breaking museum exhibits. Also housed under the roof of the new two-story addition, clad in

Dolan DNA Learning Center, main entrance.

brick and with stylistic details to match the original Georgian-Revival schoolhouse, are two pedagogically related features that couldn't be accommodated in the initial remodeling but have both proven necessary for brain refueling at different levels: a faculty conference room and student lunch room!

Besides the field trip activities, the Learning Center early on began to organize week-long summer courses for students, today as young as the fifth grade, who just couldn't get enough science – and who, in some cases, didn't realize they could become biologists or go on to obtain Master's Degrees and PhDs in the discipline. *DNA Science: A First Course*, created by DNALC staff in 1995 and now in its second edition, remains the classic high school text on the subject.

Nowadays, the general public is also welcomed in for fun-filled "Saturday DNA!" happenings. For those who cannot make it to the North Shore of Long Island, but love science, the Dolan DNALC staff have created the "Gene Almanac" Internet portal and a whole family of educational Internet sites covering broad topics, including basic heredity, genetic disorders, eugenics, the discovery of the structure of DNA, the human genome, cancer, plant genetics, and the science of the human brain. Federal and foundation grants of $25 million have helped make the Learning Center what it is today, including continuous funding since 1986 from the National Science Foundation and multiple grants from the Howard Hughes Medical Institute and the National Institutes of Health.

Besides the field trip activities, the Learning Center early on began to organize week-long summer courses for students, today as young as the fifth grade.

Landscape

A DNALC tree sampling

The entrance to the "new" Dolan DNALC is via a steep drive off Route 25A, just before the classic main façade of the former Union Free School comes into view. Today a pair of Kwanzan Cherry grace the former formal entrance to the schoolhouse and two Sawtooth Oak have also been planted at the front of the building. On top of the embankment that signals the driveway to the new entrance on the side of the facility, a trio of Japanese Cryptomeria, with their awl-like needles, enjoys pride of place; they are one of the new "compact" cultivars and so (unlike the specimens back at Bungtown) they won't become forest giants along Cold Spring Harbor's Main Street.

THE GRADUATE RESIDENCES ON HARBOR ROAD

Today Cold Spring Harbor Laboratory has its own graduate school, which operates chiefly out of the laboratories and seminar rooms along Bungtown Road. But housing is another matter! First year students in the Watson School of Biological Sciences can choose between two handsomely restored whaling-era residences, across the street from each other on Harbor Road in Cold Spring Harbor, and almost directly across the harbor from the Wawepex Building. One could row a boat from home to school and vice-versa! The graduate residence now called the **Townsend Jones Knight House** was originally built ca. 1790 for John Hewlett Jones, who, with his brother-in-law Divine Hewlett, was a partner

Townsend Jones Knight House (r), ca. 1790, and George W. Cutting House, ca. 1810. Both of these Federal-style dwellings, directly across the Inner Harbor from Bungtown, have been renovated as residences for first-year Watson School students.

in the Jones industries, which included a gristmill and a store, immediately south and north, respectively, of the homestead at the time it was built (and later woolen mills, the barrel factory, and other industries connected to whaling). The house even served as the first post office for the village of Cold Spring with Divine Hewlett serving as the first postmaster in 1826. Purchased and rehabilitated just in time to welcome the members of the very first entering class of the Watson School in the fall of 1999, historic Knight House was formally dedicated the following spring – a celebration not only of CSHL's new graduate school, but of the Laboratory's intimate connection with the Jones family of Cold Spring Harbor, its founders and nurturers for over a century.

Within a few short years, graduate accommodation jumped across the road to another building long connected to the history of Cold Spring Harbor, which had been previously acquired and renovated to house foreign post-docs without cars in close proximity to the Laboratory grounds "around the corner" on Bungtown Road. This residence, built early in the 19th century as a multiple family dwelling, was given the new name **George W. Cutting House** to mark its new life as an additional home for first-year Watson students.

Elizabeth Sloan Livingston Houses, 2005, at Agricultural Field Station. Duplex style residences for post-docs.

Landscape

New plantings

Both graduate residences on Harbor Road are in the process of having their landscapes renewed, which when completed will feature a selection of some of the best tried and true "new" shrubs and trees. A small **'Athena' Elm,** *Ulmus parvifolia* 'Athena,' was recently planted in the west lawn as a specimen tree. Because it is a patented cultivar of the Chinese, or Lacebark, Elm, "Athena" is believed to be impervious to the "Dutch Elm Disease" which affects the American Elm (*Ulmus americana*). KNIGHT is also graced with a pretty collection of Crab Apples and Cherries. Young trees situated around CUTTING include a lovely specimen of Higan Weeping Cherry, complemented by a nice representation of season-spanning, always elegant Kousa Dogwood. In addition, a cheerful planting of golden daylilies brightens its ancient retaining wall along Harbor Road.

Seed Laboratory, ca. 1920. Former main garage of Nichols estate "Uplands Farm," adaptively reused as plants genetics lab.

Research Greenhouse, 1985, adjacent to Seed Laboratory.

THE AGRICULTURAL FIELD STATION AT UPLANDS FARM

Additional graduate housing exists at the Laboratory's Agricultural Field Station at Uplands Farm, located on Lawrence Hill Road,

which veers off Route 25A in a southeasterly direction, about a tenth of a mile east of the entrance to Bungtown Road. A former Gold Coast estate, the sixty-acre Uplands Farm was donated to the Nature Conservancy upon the death of Mrs. Jane Nichols and had been in their possession for a number of years when in 1984 CSHL purchased twelve acres. Included in the Laboratory's parcel were the former main garage and two estate staff residences. The garage

William R. and Irene D. Miller Residence and (inset) Wendy Vander Poel House, ca. 1920. Colonial Revival dwellings from Nichols estate era, today housing for CSHL scientists.

structure was adaptively reused as a **Seed Laboratory,** and a 2200-square-foot **Research Greenhouse** was erected adjacent to it. The two charming staff homes on the parcel were adapted for use as graduate student housing, and are now known as the **William R. and Irene D. Miller Residence** and the **Wendy Vander Poel House.** These were supplemented in 2005 by the **Elizabeth Sloan Livingston Houses,** designed in the form of college suites housing that features individual entranceways and is clad in materials to blend with all the former estate buildings, such as clapboards and shingles, and articulated with brick foundations and chimneys.

Landscape

Trees old and new

The parcel at Uplands Farm that belongs to Cold Spring Harbor Laboratory lies immediately adjacent to the "farm square" where the Nature Conservancy has its offices and which is graced by magnificent old specimens of Sycamore, Norway Maple, and White Oak (plus a lone Pin Oak). Immediately adjacent to the "square" is a slender patch of lawn belonging to the Laboratory, on which a small **Pink Horsechestnut,** *Aesculus x carnea* 'Briotii,' transplanted from the big Bungtown field on the main campus, is now thriving. Directly behind the SEED LABORATORY you can't miss the extensive field for growing **Maize (Seed Corn),** *Zea mays,* which is used in breeding experiments in conjunction with the program in plant genetics headquartered in the Delbrück/Page Laboratory back at Bungtown. Incidentally, together with an additional smaller greenhouse, there is now enough space under glass to grow sufficient quantities of winter corn and the radish-like model plant organism, **Thale Cress,** or **"Mouse-ear Cress,"** *Arabidopsis thaliana,* for future analysis in the laboratories on the main campus.

The edge of the road into the housing area at Uplands Farm is close to the property boundary line and is lined with naturalizing Red Maple and Sugar Maple. The main façade of the LIVINGSTON HOUSES (adjacent to parking) has been planted with six small and slender trees, two Blue Atlas Cedar, two Paperbark Maple, and two **Columnar English Oak,** *Quercus robus* 'Fastigiata.' These new "houses" are screened from the back by native evergreens, such as Eastern Redcedar and White Spruce. Mature plantings from the estate era grace the older cottage-style residences.

Translational genome research and scientific publishing

In 2000 Cold Spring Harbor Laboratory opened its **Woodbury Genome Center** in the former Institute of Physics building on Sunnyside Boulevard in Woodbury, Long Island (about ten minutes by car from the Bungtown campus). The building was completely renovated and redesigned, including the addition of a striking pentagonal entrance pavilion, which has modernist stylistic echoes inside. Light-filled and airy, the facility is decorated with a plethora of photographs representing favorite places at the Laboratory. The 65,000-square-foot single-story building (with room for expansion) is housed on a twelve-acre cul-de-sac off a busy highway, with easy access to the Island's expressways and parkways. Within this stand-alone facility, four research centers

Woodbury Genome Center, 2000 - genomics facility and headquarters of the Cold Spring Harbor Laboratory Press. In foreground, Japanese Maple. Inset, enframed with 'Heritage' Birch.

173

were established: the Bioinformatics Center; the Cancer Research Center; the Genome Sequencing Center; and the Plant Genomics Center. In addition to these research laboratories, the Woodbury Center houses the headquarters of the world-renowned Cold Spring Harbor Laboratory Press, including editorial and production offices.

Genome research

In its planning stages, the mission of the Genome Center was stated in compelling and simple terms:

> The genome of every organism, its genetic blueprint or biochemical "book of life," is written within the DNA sequence of its chromosomes. An essential part of the work that will take place at the Genome Research Center will be to generate such raw DNA sequence data. However, determining the DNA sequence of individual genes, or even of an organism's entire genome, is only the beginning. To fully appreciate how such information can improve quality of life requires exploration of the deepest mysteries of gene function. This is the challenge and the hope of the new millennium. Meeting this challenge demands the development of new technologies and an inter-disciplinary approach.

While much of the sequencing effort (determining the order of the A, C, G, and T bases, or "letters," along DNA molecules) at Cold Spring Harbor used to be focused on plant DNA, today the human DNA genome (3 billion letters) has become the main focus of attention. While the United States allotted $3 billion during 1988-2003, via the Human Genome Project of the National Institutes of Health, to the international effort to work out the first human genome, continued improvements in sequencing technologies and bioinformatics may soon permit additional human genomes to be obtained for no more than several thousand dollars. These radically reduced sums will lead to a further flowering of human genetics starting with the discovery of changes in DNA sequences that give rise to individual cancers as well as many if not the majority of causes of serious diseases of the brain such as Alzheimer's, Parkinson's, autism, schizophrenia, and major depressive conditions. Already the facilities at Woodbury have positioned Cold Spring Harbor Laboratory in the forward tackle position for winning the war on all these illnesses caused by bad throws of the genetic dice.

A growing number of titles are appearing on the scene and more and more information is being published online.

Cold Spring Harbor Laboratory Press

Cold Spring Harbor Laboratory has long been in possession of a powerful vehicle for sharing the kind of information created by the researchers at the Genome Center, by those on the main Bungtown campus, and in fact by scientists all over the scientific world as well: the Cold Spring Harbor Laboratory Press. The Press relocated to Woodbury in 2001 after outgrowing its Urey Cottage home on the main campus. Now an international organization with sales and marketing offices in San Diego and the UK, and agents all over the world, the Press originated way back in 1933, with the yearly publication of the dark maroon books containing all the papers presented at each year's annual Cold Spring Harbor Symposium on Quantitative Biology.

Over time, the types of publications became more diverse: monographs were commissioned and, later, laboratory manuals based on protocols used in innovative CSHL summer courses were published to great acclaim, with Joe Sambrook's *Molecular Cloning* becoming an international and perennial bestseller. A number of research journals were founded in rapid succession (now totaling six, including two published for professional scientific societies) with an annual output of articles exceeding 12,000. A textbook division was recently established, its first publication being *Evolution*. Today there are 192 CSHL Press titles in print, including books accessible to the layman that "delve into the practice, personalities, and history of science and its influence on medicine, business, and social policy" as well as *Enjoy Your Cells, Have a Nice DNA*, and other books for children.

The Press relocated to Woodbury in 2001 after outgrowing its Urey Cottage home on the main campus. Now an international organization with sales and marketing offices in San Diego and the UK, and agents all over the world . . .

A growing number of titles are appearing on the scene and more and more information is being published online. Cold Spring Harbor Protocols is an interactive source of lab methods, and BioSupplyNet, the online laboratory supplies site long supported by the Press, now features free downloadable CSHL Press protocols for biological techniques such as PCR, RNAi, and more.

The Press alone has nearly sixty employees working out of the Woodbury Genome Center building. Considering that the daily population of the facility, including researchers and support personnel, will eventually reach a complement of more than 200 workers, it was good planning to have an

Among recent Woodbury Genome Center plantings – Ornamental Cherry along patio edge; Weeping European Filbert (still tiny, near building corner).

auditorium, dining commons, and library annex included in the
conceptual plans from the very beginning. A child care annex is
now under active consideration.

Landscape

Familiar trees

While the scope of the landscaping at the Woodbury Genome Center is not large, at least in terms of available "green space," attractive plantings featuring many of the same species seen at Bungtown (and beyond) have been installed all around the facility. The main entrance façade is graced by a pair of 'Heritage' Birch, one on each side of the entrance canopy (which just happens to have the same shape as the purine and pyrimidine bases of the DNA molecule!). Specimens of Alaska Cedar, Ornamental Cherry, Deodar Cedar, and Japanese Cryptomeria (compact form) also enliven the scene.

> . . . attractive plantings featuring many of the same species seen at Bungtown (and beyond) have been installed all around the facility.

Rounding the corner you can also see three Serbian Spruce, six Sawara Falsecypress, a planting of Redosier Dogwood, and four Bradford Pear. Behind the building, to either side of a generous patio for alfresco dining surrounded by eight Ornamental Cherry, a pair of Sawtooth Oak stand sentinel. Even the

Woodbury plantings (l-r) – Cryptomeria, 'Heritage' Birch, Deodar Cedar.

back door to the kitchen area is flanked by a pair of twisted and low **Weeping European Filbert,** *Corylus avellana* 'Pendula' (which in their habit greatly resemble the Demerec Laboratory Weeping European Beech back on Bungtown Road). And ringing the boundary fence for the facility are plenty of those old favorites, our native Black Locust, White Ash, and White Oak, plus naturalizing stands of White Pine.

En route to the new Watson School campus

As I write, six new laboratories are heading towards completion on the steep hillside above the Neurobiology Center (see Chapter Four). The focus of the Hillside Campus will be to use the new genetic knowledge to devise new cancer therapies—research which will be done in collaboration with clinicians from regional medical centers. At the top of the hill will go a purpose-built campus for the Watson School of Biological Sciences.

CSHL's future is arriving fast

Although it has given rise to six separate buildings, the Hillside Campus is in reality a single, very large construction project (much

"Hillside Campus," to be completed in 2009, incorporates (clockwise from top ctr): Wendt Family Neurosciences Building, Donald and Joan Axinn Neurosciences Building, Leslie and Regina Quick Cancer Therapy Building, William and Marjorie Matheson Cancer Research Building, DeMatteis Family Human Genetics Building, David Koch Mathematics Building. . . . A design for the "Upper Campus," future home of the Watson School of Biological Sciences, can be seen in background, at top of Bungtown hillside (watercolor by William H. Grover, Centerbrook Architects and Planners).

the way BUSH LECTURE HALL and DEMEREC LABORATORY were erected in tandem; see Chapter One). It stretches from the foot of the hill immediately behind Freeman Building, Urey Cottage, and Marks Lab, up to a plateau at the highest reaches of Bungtown. Building construction began in 2006 but site preparation was commenced in 2005; it included excavation, grading, and roadwork for construction access. Sand, gravel, and pea stone "mined" from the hillside were sold to help defray the cost of excavating the project site. In fact, it was one big hole, and, besides all being contiguous, several of the buildings are physically interconnected.

What is more, five of the six buildings can be accessed from an open-air "*Piazza*," about half-way up the hill, which is in fact the focal point of the Hillside Campus design. Three of the laboratories are situated east of, i.e. seemingly lower than, the Piazza, and the remaining three appear to rise up behind it, to the west, "behind" the Piazza. The three lower buildings are, from south to north: the **David Koch Mathematics Building;** the **DeMatteis Family Human Genetics Building,** which is interconnected to the Koch Building Laboratory; and the **William and Marjorie Matheson Cancer Research Building.** A broad stone double stairway runs upwards to the Piazza right in between the Koch and Matheson labs.

The three buildings on the west side of the Piazza are arrayed on the hillside as follows, from the north back to the south (completing a circuit around the Piazza): the **Leslie and Regina Quick Cancer Therapy Building;** the **Donald and Joan Axinn Neurosciences Building;** and the **Wendt Family Neurosciences Building,** which will also house, on the Piazza level, a lecture hall and a cafeteria. Access to the Piazza is via the lower floors of these three buildings on the west side, which are situated higher on the hill, but via the upper floors of the three buildings on the east side, which are lower down on the hill. (The lowest building, the Koch Building, has access to the Piazza via the top floor of the DeMatteis Building to which it is interconnected.)

In addition to the lower outdoor stairway which takes you up to the Piazza, there is a second stone stairway, even broader, that rises in between the Wendt and Axinn buildings, and takes you on up to the top of Bungtown. A water feature, patterned after the pebbly streams that cross over (and under!) the Laboratory grounds en route to the harbor—especially Watson Crick/Creek— will splash down the center of this upper stone stairway.

A vision for the top of Bungtown

At the top is the natural plateau that has been shaped to receive the future home of the Watson School of Biological Sciences; classrooms, seminar rooms, an auditorium, a dining room and commons, and residence halls for students and professors will be erected there. A new library will enjoy pride of place at the entrance to this complex of academic buildings, which has been anticipated ever since the founding of the Watson School in 1999. This Upper Campus is to be arrayed village-style around a "green" or "lawn" at the center, in the style of Thomas Jefferson's classic design for the University of Virginia.

Hillside Campus
construction.

Landscape

Long live the natives!

Re-planting is nearing completion in those sections of the Sugar
Maple forest community that were removed in the process of site
preparation for the Hillside and Upper campuses. Undoubtedly,
the edges of the lawn at the center of the future Upper Campus –
"Watson Yard"? – will be embellished with varieties of the most
beautiful types of specimen trees. Stay tuned for the next chapter
in the green history of CSHL!

CHECKLIST OF BUILDINGS –
COLD SPRING HARBOR LABORATORY

Checklist of CSHL Buildings, with supplementary information on architects and styles; construction history; and current use (in italics), with page numbers where discussed in Grounds for Knowledge *text. • All buildings are on the main Bungtown campus unless otherwise noted.*

Key to other CSHL Campus Sites: BC – Banbury Center, Banbury Lane, Lloyd Harbor (Huntington) • AFS – Agricultural Field Station (at Uplands Farm), Lawrence Hill Rd., Cold Spring Harbor • CSH – Route 25A, Cold Spring Harbor • WGC – Woodbury Genome Center, Sunnyside Blvd., Woodbury

AIRSLIE, 1806. Federal style residence originally built for gentleman farmer Major William Jones. Interior renovated in Post Modern style, 1974; Charles Moore Associates. *CSHL President's House,* 121

DONALD AND JOAN AXINN NEUROSCIENCES BUILDING – to be completed in 2009; Centerbrook. 182

BALLYBUNG, 1994; Centerbrook. Neo-Palladian style (symmetrical) residence, with English Regency style detailing (doors, windows, rooflines). "Ballybung" is a fanciful name connoting "place of the bung," viz. Bungtown. *CSHL Chancellor's House,* 127

ARNOLD AND MABEL BECKMAN LABORATORY, 1991; Centerbrook. Dark brick academic style, with hint of gothic buttresses. First modern facility at CSHL with purpose-built teaching laboratories. Arnold Orville Beckman (1900-2004) invented the pH meter. *Research and teaching in neuroscience,* 100

EUGENE BLACKFORD MEMORIAL HALL, 1907; Gardner & Howes. Utilitarian style. Early use of reinforced concrete for an academic facility: Dining Hall and Lounge and Women's Dormitory. Venue of first CSH Symposium in 1933. Winterized in 1973. Extensively enlarged in 1993; Centerbrook. Lobster banquets held here at conclusions of meetings and courses at CSHL. New York State Fisheries Commissioner Eugene G. Blackford (1839-1905) was a co-founder in 1890 of the Biological Laboratory. *Dining facility and dormitory,* 33

VANNEVAR BUSH LECTURE HALL,
1953; Anderson & Beckwith. Scandinavian
Modern brick and concrete structure.
Home from 1953-1985 of annual Cold
Spring Harbor Symposia on Quantitative
Biology. Renowned educator and chairman
of the Carnegie Corporation, Vannevar
Bush (1890-1974) helped found the
National Science Foundation. *Poster sessions
during professional meetings; art exhibitions,
lectures and gatherings at other times*, 32

The CABINS, 1989 and 1991;
Centerbrook. Contextual design that
resurrects the rustic style of the historic
1930s Cabins (removed in 1951 and 1987).
The 1989 assemblage, lower on the hillside,
consists of Boyer, Eagle, Glass, Luria,
Stahl, and Stent Cabins; the 1991 colony,
at the top, comprises Alumni, Maniatis,
Pall, Wendt, and Zinder Cabins. *Visitor
accommodation*, 103

JOHN CAIRNS LABORATORY
(incorporating 1910 Sheep Shed/later
called Mouse House), 1970; Charles
Moore Associates. Contextual design.
John Cairns (1922-) was director of
CSHL from 1963 to 1968. *Functioning of
chromosomes in cells*, 59

CARNEGIE BUILDING, 1905; Kirby,
Petit & Green. Renaissance Revival style.

Originally erected to be the Main Building
of the Carnegie Institution of Washington
Station for Experimental Evolution.
Headquarters of the Library since 1953.
*Archives and Information Services; new wing
which will house CSHL/Genentech Center for the
History of Molecular Biology and Biotechnology,
to be completed 2009*, 57

WILLIAM H. COLE COTTAGE, 1934.
Vernacular Depression era cottage. *Staff
housing*, 75

GEORGE W. CUTTING HOUSE, ca.
1810. Federal style dwelling first acquired
in 1989. *Residence for first-year Watson School
of Biological Sciences students (CSH)*, 167

NORRIS AND HENRIETTE
DARRELL HOUSE (incorporating ca.
1920s Ice House of de Forest estate), ca.
1950s. Modern picturesque style frame
dwelling with original stone detailing in
octagonal shape from earlier structure.
Senior staff residence, 129

CHARLES BENEDICT DAVENPORT
HOUSE, 1884. Victorian style residence.
Built for Fish Hatchery superintendent on
former site of John D. Jones house.
Became home to Charles Benedict
Davenport (1866-1944) and his family in
1904 upon founding of Station for
Experimental Evolution of the Carnegie

Institution of Washington. Then director of the Biological Laboratory of the Brooklyn Institute of Arts and Sciences, Dr. Davenport was also founding director of the Carnegie Station. House restored in 1980; paint colors show original 1884 exterior decoration scheme as determined by scientific paint analysis at National Park Service lab in Boston. Formal double parlor now Music Room with baby grand piano. *Post-doc residence,* 26

DE FOREST STABLES, 1914; Clinton McKenzie. Vernacular estate outbuilding. Ground floor renovated as Mary D. Lindsay Child Care Center, 1997; Centerbrook. *Staff apartments and childcare facility,* 124

MAX DELBRÜCK/ARTHUR W. AND WALTER H. PAGE LABORATORY (incorporates 1926 Davenport Laboratory), 1981 and 1987; Moore Grover Harper. Neo-Colonial Revival style teaching and research complex. Max Delbrück (1906-1981), founder of the CSHL Phage Course, mentored the first generation of molecular biologists, including J. D. Watson. *Plant genomics,* 79

DEMATTEIS FAMILY HUMAN GENETICS BUILDING – to be completed in 2008; Centerbrook. 182

MILISLAV DEMEREC LABORATORY, 1953; Anderson & Beckwith. Modernist concrete structure. Green aluminum wing added in 1982 and dark brick in 1989; Centerbrook. Noted bacterial geneticist Milislav Demerec (1895-1966) was longtime director, simultaneously, of Biological Laboratory and Carnegie Institution Department of Genetics. *Cancer research,* 29

CHARLES AND HELEN DOLAN HALL, 1991; Centerbrook. Light brick hotel style. First multistory handicapped-accessible accommodation at Bungtown. Charles and Helen Dolan have each served actively on the Board of CSHL. *Residence hall for visitors,* 100

DOLAN DNA LEARNING CENTER (incorporating 1925 Union Free School of Cold Spring Harbor, designed by Peabody, Wilson & Brown), renovated and enlarged, 1988 and 2001; Centerbrook. Courses on DNA for school teachers were first delivered via Vector Vans; program plus students moved into CSH schoolhouse in 1988. *DNA literacy program administration, auditorium, classrooms and labs (CSH),* 162

The FIREHOUSE, 1906. Late Victorian. Barged across harbor from CSH village to CSH Laboratory, 1930. Moved 50 yards north and renovated, 1986. *Staff apartments,* 81

SAMUEL B. FREEMAN BUILDING, 2000; Centerbrook. Contextual style facility (blends with nearby buildings). *Computational neurobiology,* 98

GALE HOUSE (incorporating early 20th century Boat House from the de Forest era); ca. 1950s. Vernacular estate outbuilding with Ranch style appendage. *Senior staff residence,* 127

GARDEN HOUSE (incorporating ca. 1920 open air wooden Garden Shelter of de Forest Formal Garden with modern brick wings), ca. 1950s. *Senior staff residence,* 129

GAZEBO, 1976; Moore Grover Harper. Post modern shingle style. Designed as secondary wastewater treatment plant; decommissioned when CSHL went onto sewer system of Nassau County in late 1980s. Won Western Red Cedar Shingle and Shake Award. *Scenic outlook,* 81

OLIVER AND LORRAINE GRACE AUDITORIUM, 1986; Centerbrook. Contextual design. Passively heated and cooled auditorium facility featuring large dormer windows, echoing those of Davenport House across road. Won ARCHI design award from LI Chapter of AIA. Meeting sessions of annual CSH Symposia and over twenty other professional meetings held here plus CSHL Cultural Series of concerts and public lectures. *Large conference facility and headquarters of Meetings & Courses and Information Technology departments,* 35

REGINALD GORDON HARRIS BUILDING, 1982; Moore Grover Harper. Contextual design. Animal facility designed along the lines of a 19th century barn that formerly stood in the vicinity. Reginald G. Harris (1898-1936) was son-in-law to Charles B. Davenport, and director of the Biological Laboratory from 1924 until his premature death. It was Harris who in 1933 founded the Cold Spring Harbor Symposia on Quantitative Biology. *Support facility,* 36

LITA ANNENBERG HAZEN TOWER, 1991; Centerbrook. Dark brick campanile style. a, g, c, and t – the symbols of the four bases that form the "steps" of the DNA "ladder" – are called out on the four sides of the Tower at its top. Its bronze bell from the Meneely foundry in Troy, NY, rings out the hours from eight in the morning till eight at night. *Academic bell tower,* 99

ALFRED DAY HERSHEY BUILDING
(incorporating 1906 Head House of former
Greenhouse complex), 1979; Moore
Grover Harper. Contextual design that
builds on foundation of former
greenhouses. Alfred Day Hershey (1908-
1997), longtime Carnegie researcher,
received a Nobel prize in 1962 for
demonstrating that DNA is the molecule of
heredity. *Offices for cancer researchers and
bioinformaticists; also Receiving Department,* 59

FRANKLIN WILLIAM HOOPER
HOUSE, ca. 1835. Late Federal style
tenement. Franklin Hooper (1851-1914), a
founder of the Biological Laboratory, had
participated in Louis Agassiz' summer
biological school off Cape Cod, MA.
Women's dormitory and staff apartments, 78

WALTER B. JAMES MEMORIAL
LABORATORY, 1929; Henry H. Saylor.
Concrete utilitarian style single-story
facility for X-ray research. Wood sheathed
2nd floor added in 1961; Geo. B. Post,
Architects. Enlarged it became headquarters for bacterial genetics and later
tumor virus research. *Cancer research,* 94

JAMES ANNEX, 1971; Edelman &
Salzman. 2nd Bay (Northern California)
wooden shed roof style. *Research staff offices,
library, and seminar room,* 94

JOHN DIVINE JONES
LABORATORY, 1893; Lindsay Watson;
remodeled with aluminum interior modules
and a wing added, 1974; Charles Moore
Associates. Colonial Revival schoolhouse
style. Won American Institute of Architects
Honor Award in Continued Use for
sympathetic adaptive reuse for
neurobiology. Originally designed for
teaching and research, it still features an
interior entirely sheathed in beaded boards
of dark-stained and varnished wood and a
huge stone fireplace opposite the front
door. John Divine Jones (1814-1895),
long-time president of the Atlantic Mutual
Insurance Company, was a founder of the
Biological Laboratory, original antecedent
to CSHL. *Research in neuroscience,* 51

TOWNSEND JONES KNIGHT
HOUSE, ca. 1790. Federal style dwelling
acquired in 1999. *Inaugural residence for first-
year Watson School of Biological Sciences
students (CSH),* 164

DAVID KOCH MATHEMATICS
BUILDING – to be completed in 2008;
Centerbrook. 182

ELIZABETH SLOAN LIVINGSTON
HOUSES, 2005; Eduardo Lacroze, AIA.
Contextual design on former grounds of
the Mrs. Jane Nichols estate, featuring

brick and shingled Colonial details. *Residences for post-docs (AFS)*, 170

DAVID AND FANNY LUKE BUILDING (incorporating 1913 Power House and ca. 1950 Carpentry Shed), 1999; Centerbrook. Contextual design that unites two formerly utilitarian structures with a tower-like monumental stair hall. David Luke is an honorary trustee and former chairman of the board of CSHL, and of the highly successful Second Century Campaign for Cold Spring Harbor Laboratory. *Departments of Public Affairs, Human Resources, and Development*, 60

EDWIN AND NANCY MARKS LABORATORY, 1999; Centerbrook. Contextual style (blends with nearby buildings). Sculpture *Twisting Dendrites* by Dale Chihuly hangs in skylit stair hall. *Neuroscience research*, 98

WILLIAM AND MARJORIE MATHESON CANCER RESEARCH BUILDING – to be completed in 2008; Centerbrook. 182

BARBARA McCLINTOCK LABORATORY (formerly the Animal House), 1914; Peabody, Wilson & Brown. Neo-Renaissance Revival style building. Substantially renovated in 1971 and renamed. Third floor added in 1987 to a

contextual design by Centerbrook based on the attic story of nearby Carnegie Building that features a monitor roof. Barbara McClintock (1902-1992), longtime Carnegie researcher, received a Nobel prize in 1983 for discovering "jumping genes." *Gene structure and function*, 58

The MEETING HOUSE (former seven-car garage of Robertson estate), 1936; Mott B. Schmidt. Renovated for adaptive reuse as meeting facility in 1977 to design of Moore Grover Harper. *Conference Room, Library, Administrative Offices (BC)*, 155

MEIER HOUSE, ca. 1970. Neo-Georgian Revival residence. *Visitor accommodation and writing center for CSHL Press authors (BC)*, 157

WILLIAM R. AND IRENE D. MILLER RESIDENCE, ca. 1920. Colonial Revival dwelling from the Mrs. Jane Nichols estate era. After leading Infrastructure Campaign, William Miller succeeded David Luke as chairman of CSHL board. *Residence for post-docs (AFS)*, 170

"NETHERMUIR," ca. 1850 de Forest family residence; demolished ca. 1943, 121

GEORGE LANE NICHOLS MEMORIAL BUILDING, 1928; Henry H. Saylor. Colonial Revival style building originally erected as a research laboratory. *CSHL central administration, 78*

The OCTAGON, 1986; Centerbrook. Post modern Victorian gazebo. Roof adaptively reused from ca. 1920 pump house across harbor. Marks halfway point on winding walk up to Neurobiology Center. *Architectural "folly," 36*

OLMSTED HOUSE, ca. 1950s. Dormered country style brick dwelling, a stone's throw from the former Tea House of the de Forest estate. *Senior staff residence, 127*

ROBERT H. P. OLNEY HOUSE, 1885. Queen Anne style residence with emblematic porte-cochere, originally to drive carriage through and drop off passengers en route to carriage house (Barn). *Post-doc accommodations, 112*

OLNEY BARN, 1885. Queen Anne style carriage house. *Grounds Department headquarters, 112*

W. J. V. OSTERHOUT COTTAGE (originally built ca. 1800), reconstructed with wing, 1969; Edelman & Salzman. Colonial style small house, modernized. *Staff housing – until Library books and staff took temporary possession in 2007, 74*

LESLIE AND REGINA QUICK CANCER THERAPY BUILDING – to be completed in 2009; Centerbrook. 182

RESEARCH GREENHOUSE, 1985. Utilitarian poly greenhouse. *Plant breeding facility (AFS), 170*

JACK RICHARDS BUILDING (incorporating ca. 1950s ranch house); renovations and additions, 1997; Centerbrook. Farm style complex with barn-like wings on north (*Mechanical Services*) and south (*Painting and Carpentry*). Jack Richards (1925-2000) was supervisor of the complete overhaul of the CSHL physical plant, starting in 1969, and director of all new construction until his retirement in 1999. *Headquarters of Facilities Department, 118*

CHARLES SAMMIS ROBERTSON HOUSE, 1936; Mott B. Schmidt. Georgian Revival style residence by renowned New York classical architect; Schmidt designed the Wagner Wing of Gracie Mansion, official residence of the Mayor of New York City. *Elegant accommodation for participants in Banbury Center meetings (BC), 155*

ROSE COTTAGE, ca. 1950. Colonial style country cottage. *Senior staff housing (BC), 157*

JOSEPH SAMBROOK LABORATORY, 1985; Centerbrook. Contextual style facility with wood cladding similar to that on 2nd floor of James Laboratory, to which it is attached. Joe Sambrook (1939-), author of the Lab's all-time bestselling lab manual *Molecular Cloning*, spearheaded tumor virus research at CSHL starting in 1969 and served as acting director of CSHL in 1983-84. *Cancer research*, 95

SAMMIS HALL, 1981; Moore Grover Harper. Post modern adaptation of Andrea Palladio's 1549 Villa Poiana (Vicenza, Italy). Symmetry is the keynote of the Palladian style on which the Georgian (and Georgian Revival) styles are based. *Residence hall for visitors (BC)*, 157

SEED LABORATORY (ca. 1920 former main garage of the Mrs. Jane Nichols estate, "Uplands Farm"), acquired and renovated in 1985. *Plant genetics and genomics (AFS)*, 170

SUPERINTENDENT'S COTTAGE, 1936; Mott B. Schmidt. Farmhouse style masonry dwelling. *Senior staff housing (BC)*, 157

TIFFANY HOUSE (incorporating late 19th century gardener's cottage), ca. 1950s. Picturesque dwelling with wing added that incorporates stylistic details from original construction. *Senior staff residence*, 125

HAROLD UREY COTTAGE, 1933; final enlargement, 1983; John M. Collins. Silk purse colonial out of sow's ear vernacular. Harold Urey (1893-1981), discoverer of "heavy water," was a Bio Lab trustee. CSHL Press first set up shop here in 1983 (now at Woodbury Genome Center). In 1999 the Watson School of Biological Sciences hung out its shingle here. *Administrative offices for WSBS and other educational programs*, 95

WENDY VANDER POEL HOUSE, ca. 1920. Colonial Revival dwelling from the Mrs. Jane Nichols estate era. *Residence for post-docs (AFS)*, 170

WAWEPEX BUILDING, ca. 1835. Vernacular warehouse style. The Wawepex Society was founded in 1895 by members of the Jones family as a holding corporation for lands which they leased to the Fish Hatchery and the Biological Laboratory. *Departments of Grants and Institutional Advancement*, 53

WENDT FAMILY NEUROSCIENCES BUILDING – to be completed in 2009; Centerbrook. 182

TIMOTHY WILLIAMS HOUSE (originally built ca. 1835), rebuilt, 1977; Moore Grover Harper. Reconstruction of late Federal style tenement – multiple family dwelling – with dormer windows and skylights added. *Bioinformatics Department; also duplex apartments for visiting staff,* 75

WOODBURY GENOME CENTER (former Institute of Physics building) extensively renovated, 2000; Centerbrook. Institutional style. *Genomics facility and headquarters and offices of the Cold Spring Harbor Laboratory Press (WGC),* 173

YELLOW HOUSE, ca. 1820. Federal style "half house" ("missing" two windows, other side of door). *Home to post-docs,* 112

BIRD CHECKLIST –
COLD SPRING HARBOR LABORATORY

Bungtown Botanical Garden – *Choose your seasons and bring your binoculars!*

Forests & Meadows

Year-Round Residents
Blue Jay
Cardinal
Carolina Wren
Chickadee
Crow
Downy Woodpecker
English Sparrow
Flicker
Goldfinch
Housefinch
Mourning Dove
Red-bellied Woodpecker
Red-breasted Nuthatch
Red-tailed Hawk
Song Sparrow
Sparrow Hawk
Starling
Tufted Titmouse
White-breasted Nuthatch

Summer Residents
Baltimore Oriole
Barn Swallow
Black and White Warbler
Bluebird
Catbird
Chipping Sparrow
Cowbird
Kildeer
Kingbird
Mockingbird
Yellow Warbler

*Migratory Visitors,
Spring and Fall*
Cedar Waxwing
Golden Crowned Kinglet
Myrtle Warbler
Palm Warbler
Ruby Crowned Kinglet

Winter Visitors
Junco
White-throated Sparrow

Snowy and American
Egrets, Inner Harbor.

Shores & Waters

Osprey leaving nest
on Sand Spit.

Year-Round Residents
Black Duck
Black-backed Gull
Canada Goose
Double-crested Cormorant
Great Blue Heron
Herring Gull
Kingfisher
Mallard
Mute Swan
Ring-billed Gull

Summer Residents
American Egret
Black-crowned Night Heron
Common Tern
Green Heron
Least Tern
Osprey
Snowy Egret
Spotted Sandpiper

Migratory Visitors, Fall and Spring
American Widgeon
Canvasback
Hooded Merganser
Northern Pintail
Red-breasted Merganser
Yellowlegs

Winter Visitors
Bufflehead
Goldeneye
Greater Scaup
Lesser Scaup
Old Squaw
Ringbilled Duck

Mute swans,
Inner Harbor.

SEASONAL INTEREST AT BUNGTOWN . . . AND BEYOND

Each season offers distinctive clues for identification via flowers, leaves, seed pods and fruits. Have fun exploring!

Identification Clues

SPRING
mid-March through mid-June

Leaf buds
Distinctive color –
Red (Red Maple)

Flower buds
Very big and distinctive
– Magnolia, Tuliptree

Young leaves
Distinctive color – Pale
red (Red Japanese
Maple), bright yellow
(Honeylocust cultivars)
Distinctive shapes –
Maple, Oak, Horse-
chestnut & Hickory, Ginkgo, Sweetgum,
Sycamore, Ash, Black Locust, Black
Walnut

Branching patterns
Still visible before complete leafing out –
See WINTER, below

Flowers – See "Flowering Sequence" below
Distinctive shape – Tulip-shape (Magnolia,
Tuliptree), candle-type (Horsechestnut,
Catalpa), Drooping (Scholartree)
Distinctive color – Bright yellow
(Corneliancherry Dogwood), green-and-
peach (Tuliptree), purple (Wisteria)

SUMMER
mid-June through mid-September
Flowers – See "Flowering Sequence" below
Distinctive shape – Drooping (Sourwood)
Distinctive color – Bright pink (Albizia),
chartreuse (Koelreuteria)

Seed containers
Samaras (Maple, Tree of Heaven,
cones, fleshy pale green upright (Cedars)

FALL
mid-September through mid-December
Leaves turning colors –
See "Fall Leaf Colors" below

Seed containers
Pods (Catalpa, Honeylocust, Scholartree),
Lanterns (Bladdernut, Koelreuteria),
Spiky balls (Horsechestnut, Sweetgum),

Nut-containing balls (Hickory, Walnut), Acorns (Oak)

WINTER
mid-December through mid-March
Please visit BUNGTOWN PINETUM,
Lower Road
Distinctive features – all trees
Bark, structure

Additional distinctive features – evergreens
Needles – shade of green, short or long, flat or rounded, bundled or single, arranged on plane or radically

Fall Leaf Colors

Yellow – Norway Maple, Ginkgo, Black Locust
Yellow/Orange/Red/Scarlet – Sugar Maple, Sweetgum, Sourwood, Sassafras, Zelkova
Maroon – Red Maple, Dogwood, Black Gum
Reddish-Brown/Mahogany – Oak
Brownish-Green/Bronze – White Ash, Sycamore

Flowering Sequence

First Appearance
March 15
Corneliancherry Dogwood – yellow
Star Magnolia – white
April 1
Kobus Magnolia – white

April 15
Bradford Pear – white
Kwanzan Cherry – deep pink
Weeping Cherry – pale pink
Yoshino Cherry – pale pink
Saucer Magnolia – white & pink
May 1
'Elizabeth' Magnolia – creamy yellow
Flowering Dogwood – white, rose
Horsechestnut – white
Crab Apples – pink
Purpleleaf Plum – magenta
Beach Plum – white
May 15
Bigleaf Magnolia – white
'Stellar Pink' Dogwood – rose
Kousa Dogwood – white
Tuliptree – green & peach
Black Walnut – white
Black Locust – white
June 1
Sweetbay Magnolia – white
June 15
Northern Catalpa – white
July 1
Linden species – yellow
Stewartia – white
July 15
Sourwood – cream
Silk Tree – shocking pink

TIMELY STROLLS – BUNGTOWN BOTANICAL GARDEN

Come and explore Bungtown – choose your season and set aside some time!

By Season

Spring – flowering trees
Bungtown Road Entrance, especially
mid-March through mid-May –
see Chapter 1
North Bungtown Road and de Forest
Drive, mid-April through mid-May –
see Chapter 5

Summer – unusual trees
Central and North Bungtown Road and de
Forest Drive – see Chapters 3 and 5

*Fall – trees with conspicuous fruits,
seedpods, and turning leaf colors*
Central and North Bungtown Road and de
Forest Drive – see Chapters 3 and 5

*Winter – trees with unusual bark
or branch structure*
Throughout Bungtown!

All Year Long
Lower Bungtown Road/CSHL Pinetum –
see Chapter 2

By Stopwatch

Quarter-hour
Trees around Grace, Blackford, Bush, and
Demerec – see Chapter 1

Half-hour
CSHL Pinetum/trees along Sea Wall –
see Chapters 2 and 3

Hour or more
Specimen trees from early 1900s or before!
– see Chapter Five

Photographers and Artists

The author is sincerely grateful for permission to include these photographs – all of them by colleagues of Cold Spring Harbor Laboratory unless otherwise noted – as well as art from other collections.

Peter Stahl – Yellow-berried Holly, balled, burlapped, and ready to move xxii • Bungtown Campus from across the harbor 13 • Eugene Blackford Memorial Hall 24 • Charles Benedict Davenport House in spring 26 • Copper Beech, Japanese Maple at Blackford Hall 40 • Tuliptree flower at Davenport House 41 • Weeping Beech at Demerec Laboratory 44 • Purpleleaf Plum, near Demerec Laboratory and Bush Lecture Hall 45 • Bradford Pear next to Grace Auditorium 46 • Scholartree at Grace Auditorium 48 • Watson Creek with bridge 50 • "Watson Crick" in winter, with White Pine 51 • Carnegie Building 56 • Hinoki cypress at Luke 61 • The Pinetum 62 • "Bungtown Pinetum" adjacent to Carnegie Building 62 • Blossom of Bigleaf Magnolia at Davenport House 65 • Sculpture *Double Strand*, by Chris Solbert, at Carnegie Building, Bottlebrush Buckeye in foreground 64 • Flowering Dogwood near Hershey Building 66 • Weeping Blue Atlas Cedar next to Jones Laboratory 69 • Sculpture *Black Oak*, by Chris Solbert, near Wawepex Building, with Ginkgo and Norway Spruce 70 • Copper Beech at Osterhout, summer 73 • Timothy Williams House and sculptures *Midnight Fair* and *Nuts & Bolts*, by Michael Malpass 75 • Red Maple near Gazebo 83 • Pussy Willow behind Delbrück/Page Laboratory 84 • American Sweetgum and sculpture *Getting There*, by Meryl Taradash, near the Gazebo 85 • Common Horsechestnut in bloom, near Firehouse 86 • Weeping Higan Cherry at Williams House 89 • Chestnut Oak and sculpture *Pyramus and Thisbe*, by Kenneth Campbell 92 • Harold Urey Cottage 95 • Edwin and Nancy Marks Laboratory, with Red Maple 98 • Beckman Laboratory 92 • Samuel B. Freeman Building with Zelkova 100 • Detail, Hazen Tower 101 • Japanese Zelkova in Neuroscience Center courtyard 105 • Sculpture *The Waltz of the Polypeptides*, by Mara Haseltine, outside Dolan Hall, with Bradford Pear 106 • Kousa Dogwood in bloom, James Annex 108 • Ginkgo near Gale House, Ballybung 110 • Seedpod, American Bladdernut near Beach Road 111 • Tiffany House 125 • Oaks at "Ballybung" 126 • Detail, entrance to Ballybung 126 • Gale House 127 • Olmsted House 128 • Darrell House 129 • Garden House 129 • Saucer Magnolia and Common Horsechestnut, Airslie 135 • Flowering Dogwood and Eastern Redcedar near Mary D. Lindsay Child

Bibliography

Barnard, Edward Sibley. *New York City Trees: A Field Guide for the Metropolitan Area*. New York: Columbia University Press, 2002. CSHL's Columnar Norway Spruce is mentioned. Great photos of mature trees.

Brockman, C. Frank. *Trees of North America: A Field Guide to the Major Native and Introduced Species North of Mexico*. New York: St. Martin's Press, 2001. Great drawings and compare/contrast notes.

Champlin, Richard L. *Trees of Newport*. Newport, Rhode Island: The Preservation Society of Newport County, 1976. This booklet describes 74 different species of trees located on estate grounds in Newport, of which many species may be found at Bungtown.

Cold Spring Harbor Laboratory Annual Reports: 1991-2006. See "Further Readings" in *Houses for Science* for citation of preceding institutional reports, dating back to 1890.

Dirr, Michael A. *Dirr's Hardy Trees and Shrubs: An Illustrated Encyclopedia*. Portland, Oregon: Timber Press, 1997. Fabulous picture book. All photos by Dr. Dirr. Many examples from University of Georgia Botanical Garden.

Dirr, Michael A. *Manual of Woody Landscape Plants: Their Identification, Ornamental Characteristics, Culture, Propagation and Uses*, Fifth Edition. Champaign, IL: Stipes Publishing, 1998. The 1187-page "Bible" of trees and shrubs.

Great Trees of Long Island. Oyster Bay, NY: Planting Fields Arboretum, 1994. Four specimens at CSHL made this list.

Hughes, Mollie K. *Forever Green: The Dartmouth College Campus — An Arboretum of Northern Trees*. Hanover, NH: The Class of 1950, Dartmouth College, 2000. Filled with interesting facts about trees in general, and certain legendary Dartmouth specimens in particular, plus locational maps galore, this book also profiles 77 different species that can be found on that tightly packed campus, of which many may be found at CSHL.

Little, Elbert L. *National Audubon Society Field Guide to North American Trees*, Eastern Region. New York: Alfred A. Knopf, 1980. Compelling (250-page) color keys of distinguishing tree features.

Long Island's Natural World. Melville, NY: Newsday Books, 2005. Entrancing drawings of our native trees.

Lotowycz, Elizabeth. "List of Woody Plant Material . . . at Long Island Biological Association." 1975 [manuscript, with covering letter from compiler to Dr. and Mrs. James D. Watson]. This is where it all began. . . .

MacKay, Robert B., Anthony K. Baker, and Carol A. Traynor, eds. *Long Island Country Houses and Their Architects, 1860-1940*. New York: W.W. Norton & Company, 1997. Source of scholarly account of Henry de Forest garden.

Murphy, Robert Cushman. *Fish-shape Paumanok: Nature and Man on Long Island*. Penrose Memorial Lecture for 1962. Great Falls, VA: Waterline Books, 1991. A moving narrative by the dean of Long Island naturalists . . . and conservationists.

Peters, George H. *The Trees of Long Island: A Short Account of their History, Distribution, Utilization, and Significance in the Development of Long Island, N.Y. . . . also a Summary of the Big Tree Census Records of 1952, 1962, and 1972, including a List of the Largest Specimens of All Species Reported*. Oyster Bay, NY: The Long Island Horticultural Society/Planting Fields Arboretum, 1973. A number of specimens at CSHL appear on the list – and still exist!

Simeone, Vincent A. *Great Flowering Landscape Shrubs*. Batavia, IL: Ball Publishing, 2005. [Photography by Bruce Curtis, Foreword by Michael A. Dirr, Introduction by Michael D. Coe.] The author is Director of Planting Fields Arboretum State Historic Park, in nearby Oyster Bay, Long Island. This small book's handsome photographs and exquisite design say "buy me and take me home with you!"

Simeone, Vincent A. "Planning and Implementation of the C.W. Post Community Arboretum." Master's thesis, College of Management, School of Public Service, Long Island University, C.W. Post Campus (Greenvale, New York), 2002. This was our roadmap for creating the Bungtown Botanical Garden.

Simeone, Vincent A. *Wonders of the Winter Landscape: Shrubs and Trees to Brighten the Cold Weather Garden*. Batavia, IL: Ball Publishing, 2005. Read this little book and become enlightened about how trees can be interesting even if they don't have any leaves (or just have needles!).

Smith, A. W. *A Gardener's Book of Plant Names: A Handbook of the Meaning and Origins of Plant Names*. New York, Evanston, and London: Harper & Row, 1963. The Latin-English dictionary for plant lovers.

Villani, Robert. *Long Island: A Natural History*. New York: Harry N. Abrams, 1997. Illuminating text and breath-taking photographs taken by the author.

Walton, Terry. *Cold Spring Harbor: Rediscovering History in Streets and Shores*. Cold Spring Harbor, NY: Whaling Museum Society, Inc., 1999. A loving portrait of Cold Spring Harbor, profusely illustrated.

Watson, Elizabeth L. *Houses for Science: A Pictorial History of Cold Spring Harbor Laboratory*. Cold Spring Harbor: CSHL Press, 1991.

Index

BUILDINGS AND TREES –
COLD SPRING HARBOR LABORATORY

About the Author

Elizabeth Lewis Watson, a native of Providence, Rhode Island, graduated from Radcliffe College and has earned two master's degrees – in Historic Preservation, from the Columbia University School of Architecture and Planning (1983); and in Library and Information Science, from the Palmer School of Long Island University (1997). She also holds honorary doctorates from the College of Charleston and Illinois Wesleyan University, where she has lectured on the preservation of historic landscapes.

Author of *Houses for Science* (a centennial history of Cold Spring Harbor Laboratory, 1991), she also drafted the nomination papers that led to placement of the Laboratory's main campus (along Bungtown Road) on the National Register of Historic Places, 1994. In addition, she authored *A Limner's View* (a sailor's view of world architecture, with "limner" Faith H. McCurdy, 1993) and contributed to *The Mansions of Long Island, 1860-1940* (1997).

Occupying various homes at "Bungtown" for nearly four decades, together with her husband James Dewey Watson and two now grown sons, Liz has always taken a delight in the grounds of the Laboratory – a pleasure heightened only recently by a course of study on the campus of the State University of New York at Farmingdale, in their program on Ornamental Horticulture. Long a trustee, plus a past president, of the Planting Fields Foundation of the Planting Fields Arboretum State Historic Park in nearby Oyster Bay, Long Island, she was instrumental in the Laboratory's joining the American Association of Botanic Gardens and Arboreta (now the American Public Gardens Association) as the Bungtown Botanical Garden in 2006.

A devoted trustee of the Society for the Preservation of Long Island Antiquities (SPLIA), Liz Watson has also served on the boards of the Cold Spring Harbor Whaling Museum and the Heckscher Museum of Art and as a member of the Huntington Historic Preservation Commission. She was appointed in 2001 to the New York State Board for Historic Preservation and currently serves also on the boards of the New York Landmarks Preservation Foundation and the Archives of American Art.

Besides having become a manic "leaf peeper" (and aspiring to "bird watcher") in her free time, Liz is an avid photographer, enthusiastic traveler, and happy hostess to those from near and far.